THE *TITANIC* EFFECT

KENNETH E. F. WATT
UNIVERSITY OF CALIFORNIA, DAVIS

THE *TITAN*

SINAUER ASSOCIATES, INC • PUBLISHERS
STAMFORD, CONNECTICUT

C EFFECT

Planning for the Unthinkable

Published by Sinauer Associates, Inc.
20 Second Street, Stamford, Connecticut 06905
ISBN: 0–87893–912–1
Library of Congress Catalog Card Number: 73–94060

This book is dedicated to
Dr. Herman E. Daly, *an economist for our future, and*
Dr. J. J. McCourt, *whose eye surgery kept me in touch with the present.*

CONTENTS

PREFACE

How does an ecologist come to write a book about economic problems? It began on a warm summer day in 1968. My wife, my two daughters and I were en route from the biggest of the Hawaiian islands to Kauai, and had broken the trip in Honolulu. After we checked into our high-rise hotel at the east end of Waikiki, I stepped out to the back balcony looking west toward Honolulu. There I saw something I had not seen previously: large numbers of tall buildings on the horizon had gantry cranes on top of them, which meant that they were still under construction. I wondered whether anyone had given thought to the number of new tourists who would be needed to fill all those rooms. I found out later that there had indeed been an overbuilding of hotel rooms in Hawaii.

This was my introduction to the phenomenon of market saturation. After that I began to notice other instances of the same thing: the empty seats in planes, the idle capacity of the world steel industry, and the young people who spent from five to eight years in preparation for a teaching career, only to discover that no more teachers were needed. I began to be very much interested in defects in the functioning of economies, and in the fall of 1968 I started collecting statistics on the subject from newspapers, magazines and government documents.

As an ecologist, I was very much aware that our society was due to run out of resources quickly, and that serious pollution problems would ultimately affect not only human health but also climate and crop growth. Now I began to realize that the problems I was noticing in the economy had the same root causes—that too much emphasis on growth, on consumption and production rather than the quality of life were leading to market saturation, unemployment, international monetary turbulence and higher taxes, as well as to pollution and resource depletion.

As I traveled abroad, other pieces of the picture fell into place. In Holland and Switzerland, I saw how much more convenient life could be in countries that rely on public transportation rather than on cars and freeways. Relatively small cities, such as Geneva, Arnhem and Beirut (all under 250,000 people) had the character of urban centers, and in all of them people predominated, not cars. I began to notice great differences in the use of land in different places. Returning to California after a trip by train from Arnhem to Amsterdam, I was struck by the contrast between the striking transition from a rural to an urban seting as the train pulled into Amsterdam, and the degree of urban sprawl as we flew south out of San Jose to Santa Barbara.

Meanwhile, more and more articles and statistics appearing in newspapers and magazines indicated serious troubles with the economy, both here and abroad. Among these had been the difficulty in selling the Concorde, the financial troubles at Lockheed, and the great losses of the airlines, particularly Pan American Airways.

Finally, I began to discover the economic function of

information. Having collected predictions by various authorities, and compared them with subsequent events, I came to see that most predictions about financial and economic matters were not very accurate. But as I watched the response of markets to these predictions, I became convinced that information about how the system is performing had become part of the system's driving mechanism. It would appear that governments are well aware of this principle, from the speeches accompanying the monthly release of statistics.

There is little reason to suppose that our society can continue much longer in its present pattern. The price mechanism does not serve to control the use of resources, nor does the government. Both may operate to prevent depletion at some time in the future, but by then it may be like locking the barn door after the horse is gone.

Recent history bears witness that a book can sometimes awaken the public to problems in need of attention before it is too late. It is my modest hope that this book may help towards that end.

Several extremely complex computer simulation models of the economy, or of the interaction between environmental and economic processes are in various stages of development. The output from these models, and the data and reasoning used in developing them, are of great importance for the future of mankind. Unfortunately, however, details of the functioning and the output from these models are difficult to communicate except to a specialist audience. This book is an attempt to present information about important processes resulting from the interaction of economic, environmental and population phenomena in a simplified, yet valid way. Instead of using complicated equations, and massive tables of statistics, arguments will be conveyed using the simplest possible tables, and appeals to the common sense and logic of the reader.

By its very nature, this is a controversial book. Many vested interests will find it to their advantage to maintain that the world's economy can survive indefinitely in its present form. The author of any book attempting to point out the built-in defects of a capitalist (or Marxist) system can expect that there will be efforts to discredit it. For this reason, special care has been taken in citing references interpolated within the text on which my assertions are based. Even so, it may appear rash to fly in the face of "experts" who have been almost unanimous in predicting continued economic growth after 1972.

In fact, little courage is required to hold an opposite point of view, once their record as prognosticators has been examined. In the December 28, 1970 issue of *Time* magazine, for example, five distinguished economists offered predictions for 1971. Three expected the unemployment rate to drop (to twelve-month averages of, respectively, 5.4, 5.5, and 5.6 per cent of the labor force) and two expected it to become stabilized at 5.8 percent, the official figure at the time they made their predictions. In fact, the figure rose, averaging 5.9 per cent of the labor force all through 1971. This was the official figure; but in fact, as this book will demonstrate, the method used by the government to compute the unemployment rate tends to minimize bad news. With certain telltale indicators included, the average rate of unemployment during 1971 amounted to 7 per cent of the labor force.

On April 21, 1971, an article in the *Wall Street Journal* contained this statement: "Stock market analysts are practically unanimous in their conviction that the Dow Jones Industrial Average will break through the formerly impenetrable barrier of 1000 sometime this year." On the day before, the average had closed at 944. It reached a peak of 950.82 on April 28, 1971. After that, it declined steadily until it reached a low of just under 800 in November, 1971. The market subsequently improved, and by January, 1973, the Dow Jones Industrial Average reached about 1060. Again the experts were euphoric; *Barron's* on January 8 announced a prediction by their expert panel that in 1973 the average would top 1200, or perhaps even 1300. In fact, the market peaked at about 1060 on January 11, three days after the article appeared, and was still dropping until the fall of 1973. Thus, twice in a row, the prediction of an upward move in the market by a consensus of financial experts was the signal for a steep decline beginning only a few days later. Evidently there is no reason to be nervous simply because of daring to contradict the experts on economic or financial matters.

Events have moved rapidly while this book was in production. We are locked into a mix of scenarios 11, 13 and 15 (Chapter 14): market saturation, resource depletion, high cost of living and dropping birth rates are operating together to produce a serious recession or a depression. The Arab nations have begun to use crude oil as a political weapon; moreover, they clearly understand the principle of speculation in oil, withholding it from the market now in order to sell it later at an inflated price. Market saturation, combined with the energy shortage, has curtailed the growth of air travel, automobile sales and housing starts (Chapter 5). The birth rate continues to drop (Chapter 7), and probably only 3.126 million people will be born in the U.S. in 1973, compared with 4.258 million in 1960. The decline in the number of births from 1960 to 1975 will probably be 31 per cent, in contrast to a 19 per cent decline from 1920 to 1935, the most serious previous drop. What will happen to pension funds when the small 1975 age class (at age 55) tries to support the large 1960 age class (at age 70) in 2030 A.D.? It is hard to reconcile that dilemma with projections of long term economic growth.

Public attitudes changed in 1973. By now, many people have adjusted to the idea that scarce resources make sustained economic growth impossible. Indeed, stopping such growth is no longer the major problem confronting mankind: it has virtually stopped. We must turn our attention to making the transition to a stable or declining economy as smooth and painless as possible.

Chapter 15 was coauthored by Dana M. Richards, based on my previous draft and suggestions by William E. Boggs and Gary Simon. Claudia Ayers wrote the material on citizens' action organizations. I am greatly indebted to Herman Daly, Marion Clawson, Andrew D. Sinauer and William E. Boggs for their suggestions after reading earlier versions of this book, and to William Murdoch and Bob Boyd for useful suggestions about particular chapters. As always, however, the final responsibility for any errors is the author's and his alone.

KENNETH E. F. WATT

Davis, California
November, 1973

THE *TITANIC* EFFECT

THE TITANIC EFFECT

THE SITUATION

The U.S. economy in 1973 had become a baffling mixture of burgeoning health on the one hand, and chronic illness on the other. Many conventional indicators of economic health were setting records. Automobile production in early March, 1973, was running about 24 per cent higher than the previous year, which had also been an outstanding one for car manufacturing. Industrial production in general was about 10 per cent higher than in 1972. Retail sales were up by about 16 per cent, electrical power production by 8 per cent. In their forecasts at the turn of the year, almost all economists and financial experts had predicted that 1973 would be a tremendous year for the U.S. economy.

In the face of such forecasts, the amount and variety of economic bad news was astounding. Indeed, the more carefully one looked, the worse that news proved to be. By March 22, confidence in the economy had so declined that of the approximately 1,800 corporations whose stocks were traded on the New York Stock Exchange, 571 were selling at a record low for 1973. In the week ending March 23, the Dow Jones industrial average for thirty key stocks

1

on the New York Exchange had declined 40.34 points, the largest drop in at least nineteen years. Inflation generally, and the rise of food prices in particular had become frightening. Unemployment was high, particularly among the young, blacks and women, as it had been for the last two and a half years. For all the surface optimism about how the unemployment problem had been controlled, joblessness among teen-agers climbed from 14.3 per cent to 15.8 per cent in the single month of January, 1973.[1]

The growth of the economy was threatened by an "energy crisis" as proved reserves of both crude oil and natural gas in the United States declined. Further economic growth had become dependent on imports of fuel from other countries whose long-term political loyalties were uncertain. International monetary turbulence had become chronic, and world prices of precious metals—the customary haven for capital when confidence in paper money erodes—had undergone wild gyrations. Conspicuous soft spots in the economy involved not only particular corporations but whole sectors of the economy. Many manufacturers of public transportation equipment and operators of public transportation systems were in precarious financial shape. Of the six largest public transportation companies in the United States—five of which had lost a great deal of money in 1970—two, Pan American Airways and the Penn Central railroad, were still losing money. Other major airlines were obviously highly vulnerable to a recession. The manufacturers of both the Concorde and the Lockheed Tristar faced an uncertain future. The world market for jet aircraft was glutted, and at home the market was becoming saturated for such diverse commodities and services as office and hotel space, toys, consumer electronics, Ph.D.'s, and elementary school teachers. Pollution was much more serious in certain urban areas than had been generally recognized, but even though knowledge of the severity of the problem was minimal, the pressure for pollution control was strong. Such control would be costly, and would impose an additional strain on the economy.

One fact had become certain. The discrepancy between the conventional belief that the economy was in robust health, and the reality that it was in trouble, had made it difficult either to explain or to predict economic behavior within the old conceptual framework, using the traditional methods. Market behavior, unemployment rates, and inflation had all become difficult to predict. The public evidently had become disillusioned with a conventional wisdom that could proclaim the economy and the market to be healthy in December, 1972, only to see the stock market decline by about 13 per cent within two months. As a result, the public appeared to be on a "buying strike" so far as stocks were concerned.

THE DIAGNOSIS

What could be the cause of this vast discrepancy between signs of health and symptoms of economic sickness? The thesis of this book is that all the symptoms, good and bad, together constitute a syndrome pointing to a single acute economic ailment: excessive, unplanned, undirected and destructive growth. By this I mean the concurrent growth in population, economic activity, the per capita use of energy and other resources, plus the resultant growth in pollution and the effects of pollution on the environment and on human health.

TWO PROBLEMS

If we are in fact confronted with a critical economic disease in urgent need of treatment, two questions need answering. How can the problem possibly be so acute when the conventional wisdom has attested to the health of the economy? Why have adequate steps to deal with the problem not been taken? The answer to the first question is to be found in the way the conventional wisdom looks at the world. The answer to the second involves two extremely important phenomena which are not widely understood: the "*Titanic* effect," and exponential growth.

THE DEFECTIVE WORLD VIEW OF THE CONVENTIONAL WISDOM

This book will argue that the conventional economic wisdom is defective in several respects, and that these defects can be remedied by incorporating into the conventional wisdom the point of view of two new sciences: ecology and systems analysis (the latter largely developed by engineers). Specifically, it would appear that the conventional wisdom focuses on measures of economic health that do not reveal the whole picture. Furthermore, economic projections currently predict the future largely by extrapolating present trends. This method of prediction is useful only for obtaining "trouble-free" scenarios, not for predicting major "systems breaks"—discontinuities caused by such novel phenomena as resource limitation, market saturation, or a slow response to crisis by a rigid bureaucracy. In short, anything that hasn't happened already is unlikely to be found in the conceptual models of the conventional wisdom.

Ecologists and systems scientists bring to the analysis of economic phenomena a number of new ideas, of which four are particularly important.

1. Everything is interrelated. This means, for example, that the U.S. economy cannot be viewed in a vacuum, but must be examined in the context of the global economy. Two typical phenomena take on a new significance when we shift to this global point of view. Although the United States does not have a shortage of food, the world does. Consequently, with the United States exporting food in a partial response to shortages elsewhere, demand for U.S. food has shot up relative to supply, and domestic food prices have risen as a consequence. Also, even though the world is not in imminent danger of running out of fossil fuel, the declining supplies in the countries with the highest use rates, such as the United States, make those countries increasingly dependent on fuel imported from the Middle East and elsewhere. This in turn has contributed to an enormous and rapidly growing trade deficit, which is largely responsible for the drop in the value of the dollar on world money markets.

2. There are limits to growth, and they are of many kinds. Economic growth is subject to limits not only on available matter and energy, but also on space, time, capital and technological know-how. Markets can be saturated, pollution expenses can become prohibitive, and there may be inadequate time to develop substitute resources. Institutions may simply balk at taking a needed step. For example, while U.S. automobile manufacturers have insisted they could not meet the proposed government requirements for controlling emission of pollutants by 1975, three foreign manufacturers have redesigned their engines so as to meet those standards.

3. Ultimately, the number of people the world can support (its total "carrying capacity") is dependent on the amount of energy put into agriculture. This input, much of it from fossil fuel, is already high and still rising throughout the world. If the supply of fossil fuel should run out before there is a substitute, the energy from the sun would be inadequate to support the population that would have been built up through dependence on fossil fuel.

4. The age structure of populations is a much more important factor in determining economic health than has been generally recognized. Much of the present unemployment is the result of a distorted age structure brought about by excessive birth rates between 1950 and 1960.

THE "*TITANIC* EFFECT"

Many people are now saying we need to do something about the "energy crisis." Yet politicians who attempt to deal with the problem are discovering that the sentiment of the electorate is against any rise in energy prices as "anti-consumer," and is also opposed to any public interest in the development of alternate energy sources at present, since that might lead to an increase in taxes. As a consequence of keeping prices down, high per capita use of energy will be encouraged and will thus continue to rise until the supply of energy runs out. And since the investment in finding substitute sources will have been inadequate, there won't be an adequate supply when we need them. In other words, the present public position is desperately short-sighted.

Yet this situation is by no means novel. History abounds with parallels of imminent disaster. Public warnings have been ignored when they were outside the range of past experience. Consequently, the appropriate countermeasures were not taken. The *Titanic* and other "unsinkable" ships that nevertheless went down; the cities built on flood plains; Pearl Harbor and other military "surprises"; hospitals and schools destroyed with great loss of life after repeated warnings of what fire or earthquake might do: these are some examples.

There appears to be a basic human tendency to ignore warnings about such possible enormous disasters as "unthinkable." We must understand this tendency and guard against it. That the world could run out of energy is "unthinkable," and consequently it is difficult to interest people in ensuring that such a thing won't happen. Yet if we examine history, an important generalization, which might be called the "*Titanic* effect," can be discerned:

THE MAGNITUDE OF DISASTERS DECREASES TO THE EXTENT THAT PEOPLE BELIEVE THAT THEY ARE POSSIBLE, AND PLAN TO PREVENT THEM, OR TO MINIMIZE THEIR EFFECTS.

In general, it is worth taking action in advance to deal with disasters. The reason is that the costs of doing so are typically inconsequential as measured against the losses that would ensue if no such action were taken.

To cure the growth disease, large steps are required. But the magnitude of the consequences that will ensue if those steps are not taken is "unthinkable." Thus, there is a real danger that nothing will be done. Remembering the *Titanic* effect, we shall be on guard against such thinking.

THE MEANING OF "EXPONENTIAL GROWTH"

"Exponential growth," synonymous with the compound interest growth of a savings account, describes the type of growth where a quantity keeps increasing by a

TABLE 1   The consequences of exponential growth

Number of years elapsed	Amount, under linear growth at 10 per cent per year	Amount, under exponential growth at 10 per cent per year
0	1.00	1.00
1	1.10	1.10
2	1.20	$1.10 \times 1.10 =$ $1.10^2 = 1.21$
3	1.30	$1.10^3 = 1.33$
4	1.40	1.46
5	1.50	1.61
6	1.60	1.77
7	1.70	1.95
8	1.80	2.14
9	1.90	2.36
10	2.00	2.59
11	2.10	2.85
12	2.20	3.14
13	2.30	3.45
14	2.40	3.80
15	2.50	4.18
16	2.60	4.59
17	2.70	5.05
18	2.80	5.56
19	2.90	6.12
20	3.00	6.72
21	3.10	7.40
22	3.20	8.14
23	3.30	8.95
24	3.40	9.85
25	3.50	10.83

fixed percentage. This is different than linear growth, where a constant amount is added each year. These two different types of growth produce startlingly different results after a very short number of years (Table 1).

TABLE 2   The number of years required for a
quantity to double at different sustained rates
of exponential growth

Per cent per annum exponential growth rate (compound interest rate)	Doubling time in years
1	69.31
2	34.66
3	23.10
4	17.33
5	13.86
7	10
10	6.93
12	5.78
14	4.95
20	3.47

Clearly, exponential growth leads to surprisingly large numbers in a short time, even when the growth rate is low. Consequently, there are no situations in which exponential growth continues for very long. Some type of saturation phenomenon (filling up all available space, or totally satisfying a market, for example) or depletion phenomenon (running out of raw materials) ultimately operates to limit growth.

Exponential growth is often described in terms of "doubling times." This is the time it takes for a quantity to double at a given sustained exponential growth rate. Table 2 gives the doubling times for various exponential growth rates. Evidently, doubling occurs in a remarkably short time when exponential growth rates are high.

One reason for the failure by our political leaders to take steps to deal with the growth disease is their lack of understanding of exponential growth.

TABLE 3   The implication of exponential growth[2]

	Number of years resource would last at an annual growth rate of:		
	3 per cent per year	5 per cent per year	7 per cent per year
Resource would last 30 years at present use rate	21	18	16
Resource would last 100 years at present use rate	46	36	30
Resource would last 300 years at present use rate	77	55	44

Once we understand the speed with which resources can be depleted under conditions of exponential growth, the urgency of taking the large steps required for dealing with the growth sickness becomes clear (Table 3).

THE PENALTY FOR CONTINUING TO FOLLOW THE CONVENTIONAL WISDOM

Our society is currently operating in accord with a defective view of reality. That is, many people assume that exponential growth in population, economic activity and resource use can go on indefinitely. The penalty for continuing to hold this assumption will be more severe in the future than it has been in the past. Formerly, a civilization that made gigantic blunders ultimately waned, to be followed by others that were more highly developed. Present-day civilization, however, is based on a technology that has become remarkably pervasive and homogeneous throughout the world. Despite superficial differences, the social and economic strategies and goals of Japan, the Soviet Union, China, England, Germany, Mexico, Nigeria, Brazil and the United States are fundamentally similar. They all allow some degree of population growth, and all promote increasing per capita use of resources. All emphasize productivity, and none emphasizes efficient use of energy. All assign a rather low priority to the quality of life, to protecting the environment, and to pollution control as compared with industrial production. This uniformity in national strategies all but ensures that a mistake occurring in one country is likewise occurring in many others. Thus, humanity has taken out no insurance against the failure of a particular set of strategies by experimenting with alternative strategies. If it ultimately appears that the social and economic policies of any one country are leading to catastrophe, the information will not be generally available much before the same problem becomes acute in other advanced countries.

A still more fundamental problem is that whereas formerly a civilization that erred would not have been powerful enough to prevent the rise of later civilizations, this is no longer true. Within the next thirty years, so much of the earth's readily available sources of energy will have been used up, in the form of oil, gas and coal, that any later society would have difficulty in attaining the present level of technological development. The decline of Rome and the ensuing Dark Ages could be followed by a Renaissance; but because the United States, Japan, the Soviet Union and other developed countries of the twentieth century are depleting the world of the fossil fuels, iron, chromium, copper, silver, nickel and many other substances that would be required to start over, a second Renaissance would no longer be possible.

Thus, the risks involved in ignoring the errors typified by the *Titanic* effect are much greater than before. The rest of this book will be an attempt to apply the concepts of ecology and of systems analysis in examining the consequences of the growth society is now experiencing.

2

THE ENERGY CRISIS

The deleterious effects of sustained exponential growth are revealed in starkest form by the energy crisis.

World supplies of fossil fuel will be gone much sooner than most people realize, if consumption continues to rise exponentially. Energy consumption abroad is rising even faster than in the United States. Governments have not been effective in regulating consumption. Because of institutional rigidity, rising prices may not lead to a sufficiently rapid development of substitute energy sources. Several real constraints, including that on the availability of capital, limit the rate of development of new sources of energy. Increased investment can accelerate the development of new technologies up to a point, but beyond that there are limits on the rate of technological innovation. Atomic energy confronts us with a host of hazards.

Clearly, then, it is not possible to deal with the energy crisis by continually trying to increase supply in order to meet demand. Rather, we must limit demand by an increasingly efficient use of energy. There are many ways in which this can be done.

TRENDS IN CONSUMPTION
RELATIVE TO AVAILABLE SUPPLY

Rather than present statistics on the depletion of a variety of resources, I shall focus on crude oil as typical, noting other resources for which the situation is similar. Table 1 gives the world production of crude oil in 1940, 1950, 1960 and 1970.[1] From these data, and from others going back to 1896, which are not included in the table, it is possible to express the rate of increase in world crude oil production, if present trends in growth of use continue, as a mathematical equation. Then, using integral calculus, cumulative estimates of world production of crude oil, up to and including each year, can be computed.

The world production of crude oil has been rising at about 6.9 per cent per year for the past seven decades. If use at this rate continues much longer, world production of crude oil within any given year will be as great as the largest new supplies being found anywhere. Exponential growth in annual production also implies a tremendous growth in cumulative production. M. King Hubbert, the most widely respected authority on world stocks of energy, places the ultimate recoverable total of crude oil in the world at somewhere between a low estimate of 1,350 billion and a high estimate of 2,100 billion barrels.[2] Critics of these estimates will need to explain why there has been

TABLE 1   Production of crude oil (billions of barrels)

	1940	1950	1960	1970	1980	1990	2000	2010
Annual	2.1	3.8	7.7	17.1	33	66	130	260
Cumulative	29	59	120	244	479	951	1,926	3,846

such a striking drop in the rate of crude oil discovery per foot of exploratory drilling. Clearly, at current rates of growth in production, the high estimate would have been passed by the year 2002, the low estimate by 1996. In short, if current trends in the growth of crude oil consumption continue, the world will have exhausted all available supplies within three decades, the exact date depending on which estimate of ultimate recoverable total is accurate.

For natural gas, the picture is remarkably similar, even for the depletion dates. Indeed, the same pattern applies to many mineral resources. If present world trends in production continue, within a few decades we shall have run out of readily available supplies of gold, silver, copper, nickel, lead, tin, zinc, mercury and chromium (an essential ingredient of stainless steel).[3] By "readily available supplies" I mean those of currently minable grades. If greater amounts of energy or money per pound of material mined could be made available, the recoverable volume of supplies would be greater. But this in turn would require the availability of more energy and more money, neither of which is a certainty.

FACTORS THAT WILL INFLUENCE THE FUTURE

For all these resources, several scenarios are possible. One is that production and consumption rates will continue to grow exponentially until they are totally depleted. Another is that growth in use rates will very soon begin to fall off sharply. Neither of these two extremes is very probable: we are more likely to experience something in between. At present, the factors determining the future use of resources seem mostly to be those that ensure an exponential growth in the rate of production. The most important of these are international patterns of development, the geographical distribution of resources, the role of advertising, government regulation, prices and substitution.

GROWTH OF RESOURCE USE
IN OTHER COUNTRIES

It is a common misconception that the United States will continue to be the dominant source of demand for resources. In fact, that demand is increasing more rapidly in other countries. This is because resource consumption per capita grows most rapidly when a country is only partly developed, before various saturation effects have begun to operate. Thus, the total demand for energy (Table 2)[4] is growing more rapidly elsewhere than in the United States; between 1967 and 1970, its share in the world total dropped from 34.9 per cent to 33.4 per cent.

TABLE 2   Increase in consumption of energy

	Total consumption of energy in billions of short tons of coal equivalents		*Average annual per cent change*
	1967	*1970*	*1967 to 1970*
World	6.1887	7.5433	6.82
U.S.	2.1571	2.5158	5.26
World Less U.S.	4.0316	5.0275	7.64
U.S. As Per Cent of Total	34.9%	33.4%	

The figures in Table 2 provide clues about the future, and also suggest an important generalization about the dynamics of economic development. Total consumption is rising faster elsewhere than in the United States because consumption per capita grows at a rate determined by the level of consumption per capita. For countries such as India and Pakistan, which are in the very early stages of development, consumption of energy per capita is increasing at only about 2 per cent per annum. In countries that have reached an intermediate stage of development, such as Japan, Greece and Italy, where the consumption of energy per capita amounts to the equivalent of anywhere from 1,500 to 6,000 pounds of coal per year, the consumption of energy increases at up to 14 per cent per annum. However, by the time a country is using the equivalent of 6,000 pounds of coal per capita in a year, the economy has matured, and there is a decline in the annual per capita growth in energy consumption.[5] This is the current level in the U.S., Canada and several European countries.

From all this it is clear that many other countries whose economies are now growing at a faster rate than our own can be expected to begin overtaking us in their total consumption of energy. Consequently, any planning based on the expectation of meeting domestic deficiencies in raw materials by increasing imports must be examined carefully. Increasingly from now on, materials we may wish to import will already have been imported or used by some other country.

There is a further complication in that the vast majority of the world's population have not yet begun to consume resources at a very high rate per capita. What will happen if India, Nigeria, Mainland China, Indonesia and Syria, among many others, begin to consume energy and other resources at the same per capita rate as the United States?

Extrapolations from available data suggest that the rate of increase in the per capita demand for resources will not reach its peak before the year 2010. Whether the less developed countries will ever be able to catch up depends on how many of the world's resources will have been used up by that time. Obviously, a Pandora's box of problems has been opened up by promoting abroad the idea that everyone should live the way we do. Our intent has been to open up markets for our products, but the effect in the long run will be to intensify competition for the raw materials we use in manufacturing and transportation. By creating an insatiable demand for goods in the Third World, we have in effect exported our economic follies.

Consumption per capita increased at a very high rate in the United States between 1890 and 1910, when the population totaled about 80 million. Now that we are advertising our life style to the rest of the world, however, a combined population of two billion will be increasing its per capita consumption at a rapid rate, if this transition to an industrialized society is ever to occur.

Will the world's supplies of crude oil hold out long enough so that India can make use of them at the rate per capita we now use them? To arrive at an answer, two projected trends must be compared: the cumulative world-wide consumption of crude oil, and the world-wide average consumption per capita. By 1996 the projected average per capita consumption is 16.5 barrels, the figure for the United States in 1960. This means that if the consumption of crude oil continues to rise at present rates, the supply will have run out before the per capita consumption throughout the world is as high as it was in the U.S. during the early 1970s. As a result, countries like India will never have the opportunity to equal our use of fossil fuels.

By advertising the U.S. life style to the rest of the world as a desirable norm, we are thus holding out goals that are unattainable for much of the world's population. Frustrating the ambitions of very large numbers of people is a practice whose consequences are likely to prove calamitous (consider, for example, the French and Russian Revolutions).

There is a further complication. In 1970, the United States used about 4.0 billion barrels of crude oil. Proved reserves in 1971 were only 38 billion barrels,[6] which could be expected to last less than a decade. Thus, the United States is utterly dependent on importing crude oil from countries where reserves are many times greater. Will those countries always see fit to export their supplies? What will we sell on world markets to raise the money for the purchase?

THE ROLE OF GOVERNMENT AS A REGULATORY AGENCY

Ideally, government should act to prevent a social system from getting out of control. When it comes to the depletion of resources, however, government has shown an alarming tendency to do just the opposite. Because the government is highly sensitive to pressure from vested interests, and the electorate as a whole is not well informed, the government follows policies that encourage rather than discourage the rapid depletion of the world's stocks of energy. Thus, the Price Commission keeps energy prices down at a time when allowing them to rise would discourage the inefficient use of energy, and when strict import quotas on oil and gas would have the same result. Instead of promoting greater efficiency, and the development of substitute sources such as solar power, the government is allowing tremendous increases in energy imports, thus missing an important opportunity to regulate the use of energy. Over the short term, it is apparently futile to expect the government to act as a regulatory agency. As the public arrives at a clearer understanding of which policies are best over the long term, however, we can expect to see officials elected with a real mandate to regulate the use of resources.

THE IMPACT OF RISING PRICES
ON ENERGY USE

In economics, an article of the conventional wisdom is that as a resource becomes scarce, the price rises accordingly and there is a decline in the use of that resource. Unfortunately, the historical record does not bear out this assertion. The exceptions range all the way from seafood to electricity. They include, on the one hand, those cases where price increases have not been sufficient to discourage use, for the reason that incomes were rising at about the same rate; on the other, and still worse, are those cases where use has gone up but prices have not. In 1960, 234 million pounds of tuna were imported into the U.S. at a price of 12.6 cents a pound; by 1971, with the price at 23.5 cents a pound, the total imported had risen to 473 million pounds.[7] The consumption of mercury, as Thomas Lovering points out, seems to have risen along with the price.[8] The use of electricity in the United States is rising about 8 per cent per year, yet there was virtually no increase in price from 1950 to 1970. If the prices of electricity and of critical raw materials such as oil and gas were to go up, the firms making use of them would have to use them more efficiently. It is at this point that the government should concentrate its limited regulatory ability.

WILL TECHNOLOGICAL INNOVATION SAVE US?

It is widely believed that as energy or other critical raw materials become scarce, the problems of supply will be solved by using alternate energy sources, such as atomic or solar energy. This theory will hold only if there is adequate investment in developing the alternate sources while oil, gas and coal are running out—which, in turn, will happen only if the prices of currently used depletable resources are kept high enough to finance development of the substitute.

It cannot be assumed that technological innovation will always occur in the nick of time. Major technological innovations generally require a great deal of capital, invested over a long period during which there is no return on investment; and success does not follow a fixed schedule. The belief, for example, that a breakthrough in developing a nuclear fusion reactor is imminent has been widely held for thirty years, yet in the early 1970s the availability of nuclear fusion as a commercial source of power was still at least four decades away. Moreover, once the first prototype plant becomes operational, building large numbers of power generating plants of novel design will mean an additional delay of six years at a minimum, even under emergency conditions, from the time the contracts are let until the plants are completed and delivering electricity into the national grid. But if crude oil and natural gas were to run out, say, in the decade between 2000 and 2010, it would be necessary not just to build a few fusion and breeder generating plants, but to replace the entire national energy generating system with a new one. How long would that take? The answer suggested by past events is not very reassuring.

Crude oil and natural gas were clearly recognized as potential sources of energy by 1865. But for oil it was not until 1925, and for gas not until 1955, that as much as 10 per cent of the world's total was being supplied by these sources of energy—a time lag of sixty and ninety years respectively. To reduce this lag to as little as ten years would require a tremendous ability on the part of all our major institutions to respond swiftly in a period of crisis. The judgment as to whether government and industry in the U.S. are capable of that response is for the reader to make. But when automobile manufacturers assert that they cannot produce smog-free cars in five years (a minor innovation as compared to breeder or fusion reactors), one is hardly reassured.

If the supply of oil and gas were suddenly to run out, nuclear plants might well be called on to supply, within ten years, not 10 per cent but more like 70 per cent of the nation's energy. Such projections argue strongly for an economy with lower energy requirements rather than one calling for breeder reactors and plutonium, with its multiple attendant hazards.

If crude oil and natural gas are to run out in the next few decades, it is important to postpone rather than hasten the depletion dates, so as to allow more time for a conversion to the use of alternate energy sources. As things stand at present, fossil fuel companies do not find it in their own interest to make their stocks last. Oil and gas companies have no purely economic motive for making their stocks last as long as possible, since their ultimate concern is with the increase of capital, the rate of increase in earnings per share, their profitability, and the share of the market they capture. The faster they deplete the world supply, the sooner they will be able to invest their capital in new ventures. There is no clearer proof of these statements than a look at the advertising of oil and gas companies, which is obviously designed to encourage people to use up the world stock of fossil fuels as quickly as possible. This picture would change considerably, however, if the government were to begin allowing significant increases in energy prices from now on. The present low prices of energy conspire to shorten rather than to lengthen the time left for developing substitute energy sources.

It is not true that the rate of research and development can be speeded up indefinitely as more capital is invested for that purpose. Although a great deal has been invested in research on cancer, on fusion energy and on defensive missiles, to cite three examples, the results have been unsatisfactory. And once the research and development of an alternative system of providing energy has been completed successfully, the amount of capital needed to implement it might exceed the supply of capital available.

CAPITAL AS A CRITICAL LIMITING RESOURCE

The capital required to convert the U.S. energy system to one based on nuclear power can be calculated as follows: In 1971, about 30 per cent of the energy consumed in the United States was supplied by electricity,[9] drawing on an installed capacity of about 387 thousand megawatts.[10] If, in the year 2000, half of U.S. energy needs, or an estimated 150,000 trillion British Thermal Units[11], were drawn from nuclear-powered electrical generating plants, a total generating capacity of 3396 thousand megawatts would be required.[12] Recent calculations by the Atomic Energy Commission put the cost of new generating capacity in the form of nuclear plants at about $20 billion for 160 thousand megawatts.[13] Thus, the capital requirement to build enough generating plants to supply half of the U.S. energy needs in the year 2000 would be $425 billion—i.e., 3396/160 × $20 billion. In 1971 the entire gross national product came to $1,047 billion. Assuming a rise by the year 2000 to $2,312 billion at the 1971 valuation of the dollar, 425/2312 or 18 per cent of one year's gross national product would be required to install the needed power capacity. As of 1971, the value of the entire U.S. electrical utility plant stood at $93 billion.[14] The building of this plant, of course, has taken place over a period of many years.

If construction of this sort suddenly became necessary in order to maintain an adequate source of electrical energy for the U.S., capital markets might well be strained to the limit, even assuming that the capital cost of a power plant would remain constant in the face of a vast increase in demand for capital. Interest rates would rise sharply and bond ratings would drop, thus limiting new investment across the board. Ordinary citizens would be discouraged by the ensuing high interest rates from owning their own homes, and it is doubtful whether a continued supply of "cheap" energy for frivolous and inefficient uses would appear worth the price.

Thus, it appears not only that to speed up technological innovation indefinitely, simply by increasing capital input, may not be possible, but also that even if it were, sooner or later there would probably not be enough capital to sustain the venture.

The intensive demands made on capital by the

nuclear generating industry have already been noticed by businessmen, and have led them to some conclusions that are of extreme social importance, as well as historically novel. The immense capital demands placed on industry for the generation of nuclear power may be such that even the largest corporations in the country can get into the business only by combining in groups. For example, the possibility of constructing a uranium enrichment plant—at a recently estimated cost of $1.5 billion, not including the cost of the power station that would be required to run the plant—is "being investigated" by a consortium of domestic and foreign corporations, including Bechtel, Westinghouse (the fourteenth largest corporation in the U.S.) and Union Carbide (the twenty-fifth largest).[15] We are presented with the amazing prospect of a business venture "absolutely essential to national survival" but too expensive for any corporation, or possibly even any group of corporations, to undertake.

THE CAPITAL HAZARDS OF A HIGH RATE OF TECHNOLOGICAL INNOVATION

In the preceding discussion the financial hazards of the nuclear power industry have in fact been underestimated. Its most serious problem is technological and economic uncertainty, entailing massive financial risks because of the high rate of technological obsolescence. Contributing to the uncertain climate for investment in nuclear power is the variety of reactors that can be built.

The first of these is the burner fission reactor, in which heavy isotopes emit radiation while fissioning into lighter isotopes, thus generating energy. Such reactors burn up the fissionable material just as if it were coal, with the result that it is of no further use in generating energy.

A second type, the converter fission reactor, produces as a side product some material useful in subsequent fission reactions. By using reactors of this kind, the total world supply of uranium oxide, the original fissionable raw material, can be made to last longer.

A third basic type is the breeder fission reactor, which theoretically can produce a large quantity of fissionable material per unit of uranium used in obtaining energy, thus greatly extending the usefulness of world stocks of uranium as compared with burner and converter reactors. Indeed, if the consumption of energy continues to grow at present rates, burner and converter reactors will have used up all the cheap world supplies of uranium in about four decades. Thus far, however, breeder reactors have not fulfilled their main promise, that they would buy us the time in which to develop fusion reactors. The prototype breeder reactor, the Enrico Fermi plant near Detroit, suffered a partial core meltdown in 1966 and in the entire ten years since its completion has produced very little electricity.

Associated with the breeder reactor are at least three major hazards: first, the possibility of its getting out of control and releasing poisonous material in the environment; second, the problem of disposing of the waste material, which must be kept away from anything living for thousands of years; third, and possibly the most serious of all, that plutonium, an ingredient of atomic bombs, is a by-product of breeder reactors. Since, as recent events have shown, guerrilla activity of an insanely suicidal sort cannot be ruled out as a possibility, there is a real danger that organized brigands might raid a breeder reactor, either to get the material for home-made atomic weapons or as a form of political blackmail. Clearly, unprecedented security around all breeder stations, and in connection with the material shipped out of them as waste, would be mandatory.

Such dangers are in fact so great that breeder reactors would appear to entail excessive hazards as a source of energy. Far preferable would be a strategy cutting back the use of energy through increased efficiency, and relying on coal and solar power until less hazardous sources of nuclear power can be developed.

The advantage of a reactor that derives energy from the fusion rather than the fission of atoms—as when deuterium, an isotope of hydrogen, combines with its fellows to form an isotope of helium, for example—is that the raw material, hydrogen, is the ninth most common element on earth. The chief obstacle to its development is that the fusion process occurs only under conditions of intense temperature, pressure and radiation. It has not yet been possible to obtain large amounts of energy from fusion except in the wildly unstable environment of the hydrogen bomb. Scientists and engineers are still trying to find a means, perhaps by using lasers or magnetic fields, to create an environment in which a controlled fusion reaction will persist.

The relative lack of public concern about the "energy crisis" is probably due to the popular belief that an infinite supply of energy has already been assured by the discovery of atomic energy. This distorts and simplifies the actual situation. Only controlled burner and converter reactors have been made commercially useful. Breeder reactors are not yet available commercially and it will be a long time before the commercial use of fusion reactors becomes feasible, if it ever does.

But for business executives who must decide on a

particular nuclear power generating process in which to invest their corporation's capital, more than a selection from among the four basic types is required. Each of the four has several variants, for example, in the media used for reducing the extreme heat generated at the core of a nuclear reactor: these include boiling water, gas and liquid sodium. The businessman who chooses among such alternatives must do so knowing that in a few years any choice he makes may have been rendered obsolete by some unforeseen new discovery. He is thus confronted with a harrowing situation: on one hand, the cost of entrance is higher than anything corporations have ever contemplated before; on the other hand, the risks of making a bad decision are also high. The amount of capital required is so large that a "bad" decision could wipe out his company. For example, a subsequent technological breakthrough could make it possible for competing corporations to enter the field with a new type of plant that could deliver a kilowatt hour of electricity at a lower cost. It is thus conceivable that the original plant might have no more than a few years of useful life before it became noncompetitive. There would simply not be enough time in which to depreciate such massively expensive capital ventures. Faced with this prospect, the decision-maker might opt for an investment in coal-fired generators or for no investment at all. The point at which an executive concludes that nuclear energy is a risk worth taking thus may come too late for the requirements of society. As a consequence, society is confronted with a dilemma: we do not want government to have a monopoly on nuclear energy generation, because of the implicit concentration of political power; but it may be that government alone can provide the necessary capital.

This is indeed the heart of the "energy crisis": that any solution which assumes a constantly increasing supply must come at so high a price. For example, the cost of a natural gas pipeline from northern Canada and Alaska to the forty-eight coterminous states has been put at five billion dollars—an amount that "would strain the limits of North America money markets and could be done at all only if utility commissions and consumers will accept the high prices."[16] And since it may prove impossible, as has already been pointed out, to permit energy use per capita to go on rising, the only rational approach to the "energy crisis" may be to begin curtailing the increase in consumption.

INCREASING EFFICIENCY IN USE OF ENERGY

To deal with the energy crisis without any decline in the quality of life, but rather with the effect of improving that quality, all we need to do is stop thinking about how to get more energy, and to focus instead on how to make better use of the energy we have. It turns out that the possibility of improving our use of resources is very great, given the awesome degree of inefficiency in present-day technological societies. In addition to cutting down on the wasteful expenditure of energy and the attendant pollution, the present inefficiency in the use of land could be greatly reduced.

Inefficiency in the use of resources runs throughout present-day technological society. It is found in the excessive dependence on electrical appliances in the home, in the planned obsolescence of those appliances, and the failure not only to use rooftop solar energy collectors as a means of regulating temperatures in the home, but also to ensure proper insulation of homes—by the use, for example, of double windows with a tightly sealed space between panes, and of vestibules to cut down the heat exchange between buildings and the outside.

But it is in passenger transportation that this inefficiency, coupled with other economic follies, becomes truly monumental. In 1969 the volume of traffic between cities, in passenger miles, was as follows: private automobiles accounted for 86.5 per cent, airways for 9.8 per cent, buses for 2.2 per cent, and railroads for 1.1 per cent.[17] The efficiency of these modes of transportation in the use of fuel is almost the reverse: private automobiles average 18 passenger seat miles per gallon; airways, 13; buses (assuming ten passengers to a ride), 54; and electric trains, 1,130![18] Railroads use far less fuel to perform a passenger mile of work, and since pollution is related to the amount of fuel used per passenger mile, it is almost certainly less. Railroads also use far less land for a given volume of traffic than other modes of transportation. The numbers of persons who can be moved per hour by different modes of transportation through a one-foot width of right of way are as follows: for cars averaging 1.5 passengers each, in an urban street 24 feet wide, with mixed traffic, 38 at a speed averaging 15 miles per hour; for buses with 32 passengers, the number goes up to 372 at a speed of 8.6 miles per hour; for an urban railway line, it rises to 2,900 persons at a speed of 18 miles per hour.[19] Thus, an urban railway line uses land about seventy-seven times more efficiently than cars on a highway, and moves people faster. (Cars can move faster if the width of the right of way is increased, but their efficiency in the use of land cannot begin to match that of railways.)

The argument in America against using trains has been that they are slow, inconvenient and uneconomical—

although in fact trains in many other countries are cheap, convenient and generally the fastest means of portal-to-portal transportation on intercity routes. On the highway parallel to the Geneva–Lausanne railway, many cars go very fast; yet I never saw a car pass the train. In Japan, the Tokaido high-speed express travels between Tokyo and Osaka at speeds up to 175 miles an hour. As a replacement for the wheels-on-rails support systems of the past, Hitachi is developing a magnetic floating system in which repelling magnets are used instead of flotation by compressed air. The result is expected to be a train that is comparatively silent, and capable of speeds up to 310 miles an hour. High-speed trains traveling from the core of one city to the core of another would be a tremendous improvement on the current system of intercity travel in which masses of cars can be trapped in 300-mile-long traffic jams, and in which jets land a one-hour drive out of town. Technologically novel train systems could be made still more convenient by sophisticated feeder systems that would meet the trains and transport passengers directly to the block they live on.

The controversy over whether the Boeing Company should be given $250 million to continue work on supersonic transports offered an opportunity for the government to play an important role in shifting economic patterns of U.S. development; for example, Boeing and other manufacturers might have been given $250 million each for conversion to research, development and the eventual manufacture of such trains. Instead, the emphasis on air travel has continued in the face of growing economic difficulties (Chapter 5).

If the rapid growth of consumption continues in only the developed countries, the drain on world energy resources over the next few decades will be severe. Since the same pattern is now spreading throughout the world, however, the problem is all the more acute. For several reasons, it is unlikely that atomic energy can soon replace fossil fuels. Consequently, runaway growth in domestic and world-wide use of energy must be stopped. A painless way of doing this, with many side benefits—including less congested traffic, less pollution, and lower energy prices—is to make more efficient use of energy.

3

THE RISING PRICE OF FOOD

Until 1972, food prices in the United States rose only slightly from year to year. In a country blessed with enormous tracts of productive land and an advanced agricultural technology, with a relatively low population density, there had thus far been no difficulty in supplying the demand for food. But understandings with Russia and China arrived at by Nixon and Kissinger in 1972 had a profound effect on international trade. Whereas previously there had been relatively little trade between certain economic blocs, those understandings led to something much more like a global trading system, in which any nation with a shortage can meet that shortage with purchases from any other nation. As a consequence, the price of food here will increasingly reflect global rather than national trends in supply and demand. From now on, because the global food supply is not keeping pace with the increasing demand, and the United States will be selling massive amounts of food abroad, prices here will increase rapidly.

Two considerations make it appear likely that this situation will not stabilize, but rather will grow steadily worse. First, the world-wide supply of food is almost certain to grow more slowly than the demand. Whereas the population of the globe is increasing at a rate of 2 per cent each year (doubling time of only 34.66 years), the rate of increase in food production is being affected by several factors: a worsening of weather conditions in the Temperate Zone brought about by increased pollution of the air; a reduction in the supply of agriculturally productive land as a result of urbanization, including highway and airport construction; and the degradation of the soil. Yet another factor, and one of increasing importance, is the rising cost of energy, which is the basis for all agricultural technology—including farm mechanization, the production of fertilizer and pesticide, and such processes of food handling as the drying of damp grain.

A second consideration is the prospect that the United States will never again be in a position to withhold grain from countries that need it. As more and more countries are able to compete by manufacturing the same products and selling them for less, the U.S. will be forced to sell increasing quantities of food to other countries simply to raise the funds for the purchase of oil and gas, our own supplies of which will soon be gone. It is thus inevitable that rapidly increasing food prices will continue to be a fact of life in the United States for some time.

WORLD-WIDE FOOD SUPPLY AND DEMAND

What is now taking place is clear from recent figures on world-wide supply and demand for rice, wheat and meat (Table 1).[1] Only potatoes, corn and fish are of comparable importance, and for these the trend is similar. World-wide food production per capita rose slightly from 1960 to 1968, but has since declined. This led to the enormous demand for U.S. grain that became evident in 1972. The extent of that demand is likely to surprise most

TABLE 1   World population and food production

	1960	1968	1970	Annual percentage growth[a]		
				1960–68	1968–70	1960–70
Population (Millions)	2,982	3,490	3,632	1.99	2.01	1.99
Rice (Millions of Short Tons)[b]	261	314	337	2.34	3.60	2.59
Wheat (Millions of Short Tons)	269	366	349	3.92	–2.35	2.64
Rice and Wheat (Millions of Short Tons)	530	680	686	3.16	0.44	2.61
Rice and Wheat Per Capita (Tons)	.178	.195	.189	1.15	–1.55	0.60
Meat (Millions of Short Tons)	68	86	88	2.98	1.16	2.61
Meat Per Capita (Tons)	.023	.025	.024	1.05	–.02	0.43

[a]Minus signs indicate a decline in the growth rate.
[b]A short ton is 2,000 pounds; a long ton is 2,240 pounds; a metric ton is 2,205 pounds.

readers. The increase in total production of wheat and rice in the United States from 1960 to 1971 (Table 2)[2] has not been spectacular. That neither the acreage under cultivation nor the yield per acre has risen greatly becomes sig-

TABLE 2   U.S. production of wheat and rice

	1960	*1971*
Wheat:		
Millions of Acres Harvested	51.9	48.5
Millions of Bushels Produced	1,355.	1,640.
Millions of Short Tons Produced	40.7	49.2
Yield Per Acre (Bushels)	26.1	33.8
Rice:		
Millions of Acres Harvested	1.595	1.818
Millions of Short Tons Produced	2.75	4.20
Yield Per Acre (Pounds)	3,423.	4,638.

nificant when the magnitude of cereal grain shortages throughout the world is considered. Wheat yields, for example, rose from 26.1 to 33.8 bushels per acre over that eleven-year interval—an average increase of 2.3 per cent per annum. From this it may be inferred that to suddenly double U.S. wheat production—as would seem to be required to deal with the current international shortage— would not be easy.

In 1972, the Soviet Union, China, India, Pakistan and Indonesia all experienced critical shortages of cereal grain. By the end of that year, the Soviet Union alone had purchased 400 million bushels of wheat from the United States,[3] and altogether the sales of wheat by the United States to other countries amounted to 1.1 billion bushels.[4] But in 1972 the U.S. grew only 1.6 billion bushels of wheat. Recent sales to other countries thus amounted to roughly 71 per cent of the entire 1972 crop! The clear inference must be that the world carryover or reserve supply of grain is being reduced. The global effect of another season of bad weather, such as occurred in the Soviet Union and elsewhere in 1972, on food reserves could be extremely serious.

The rising demand for agricultural commodities is having a dramatic effect on wholesale prices. From 1946 to 1971, the annual averages for the wholesale price of wheat, for example, fluctuated between $1.24 and $2.07 per bushel. The price at the end of June, 1972, was $1.50. From then until January, 1973, the price rose almost continually, to $2.72 per bushel. With the prospect that bad weather in 1973 might lead to a further severe grain shortage in the Soviet Union, there was little reason to hope for much slackening of the pressure on food prices in the U.S.

FACTORS DETERMINING AGRICULTURAL PRODUCTION

The price of food in the future will depend on supply and demand. The demand per capita can be estimated from a knowledge of nutritional requirements. The supply of food can be computed from the number of acres available, and the yield per acre—which in turn depends on the strains of food grown, the input of technology (mechanization, fertilizer, pesticides, etc.) and the weather. Fortunately, a great deal is known about all of these. The total acreage of food-growing land has been estimated very precisely, and we know the rate at which this acreage is being lost to urbanization, and to other forms of land use, per person added to the population.

At the present rate, the world population after a few decades, will be made up of about 42 per cent children under fifteen, and 29 per cent each of women and men fifteen years of age and older. The nutritional requirements per capita for each of these population groups are as follows: for children under fifteen (42 per cent of the population), 2,000 Calories; for women over fourteen (29 per cent of the population), 2,400 Calories; for men over fourteen (29 per cent of the population), 3,000 Calories.

The weighted mean population food requirement, world-wide, thus comes to 878,190 Calories per year.* The total world food requirement for any year in the future, in Calories, can be obtained by multiplying this number by the projected world population for that year.

An estimate of the total world acreage available for crop production made by the Dutch agronomist, C. T. de Wit,[5] who examined each ten-degree latitudinal belt of the earth for potential crop-growing surface, puts the global total at 5.65 billion acres. This total is continually being subtracted from, however, as land is used for cities, highways, airports or other non-agricultural purposes. Once again, careful estimates have been made of the land required for such purposes, per person added to the population.[6] For the United States, these estimates are as follows: for urban development, .208 acres; for highways and airports, .079 acres; for reservoirs and flood control, .208 acres; for wilderness parks, recreation and wildlife, .492, adding up to a total of .987 acres.

Of this .987 acres of land per person added to the population, however, only the .287 acre for urban development, highways and airports is subtracted from land flat enough to be potentially useful for agriculture. In most countries, moreover, agricultural land is already in such short supply that the immense wastage implicit in urban sprawl cannot be tolerated. In Europe, for example, the

*$(2000 \times 42) + (2400 \times 29) + (3000 \times 29) = 2406$ Calories per day.

extra land per person added to the population devoted to urban uses would only be about .185 acres.[7] In densely populated cities, such as Manhattan or San Francisco, the total land taken up per person, including highways and airports, is only about .064 acres.

In estimating the amount of food that can be grown per acre, it is of interest first of all to know how great an improvement in yield can be attained by a truly heroic effort to use new, improved strains, coupled with fertilizer, mechanization, etc. The most striking instance of which I am aware occurred in Mexico,[8] where the average wheat yield between 1948 and 1952 was 880 kilograms per hectare. By 1967, the yield had been increased to 2700 kg./ha., an average annual improvement of 6.8 per cent. It is doubtful that crop yields anywhere can be improved at a faster rate over a period of two or more decades. That this per annum rate of improvement is relatively small suggests the magnitude of the problem of the international shortage of wheat that became acute in 1972.

Some readers may find this rather sober assessment difficult to reconcile with their impressions of almost miraculous agricultural improvements supposedly attainable with the much-advertised "Green Revolution." The facts of world food production are recorded in Table 1. There are several reasons for these facts, which indicate that the Green Revolution has been less than miraculous. The new miracle strains of wheat and rice require heavy reliance on inputs of agricultural technology: water, machinery, fertilizer and pesticides. Farmers in desperately poor countries may not be able to afford these inputs. Energy with which to provide the inputs is actually more expensive in India than in many much more affluent countries.

As has been noted, agricultural production is dependent on mechanization, fertilizers and pesticides, which in turn require an enormous input of energy from fossil fuel. Furthermore, agricultural production is notably subject to the law of diminishing returns, as is demonstrated by a comparison of figures on the average yield of major crops for three countries in 1963, and on the approximate amounts of fertilizer applied in the same year. In India, the yield was 820 kilograms per hectare, with an application of approximately 2.6 kilograms of fertilizer; for the U.S. the yield was 2,600 kg., the application of fertilizer approximately 47 kg.; for Japan, the corresponding figures were a yield of 5,480 kg., and an application of 280 kg. per hectare.[9] To bring the yield for India up to that for U.S., an increase in yield of 3.2 times, the application of fertilizer would have to be increased 18 times over. To bring the yield for the U.S. up to that of Japan, a further increase of 2.1 times, the application of fertilizer would have to be increased about 6 times over. To increase the yield for the U.S. 3.2 times over, to 8,320 kilograms per hectare, the application of fertilizer would have to be increased to 1,1110 kg. per hectare, or 23.6 times over.

Clearly the cost of these incremental additions of farm technology, all of which represent energy inputs, is such as to impose severe limits on agricultural productivity.*

---

*The sensitivity of agriculture to different factors may be stated in the following formula:

Let N represent the number of people living in the world in 1970 (3.632 billions);

X represent the number of people that could be added to the 1970 population, in billions;

A_f represent the amount of land available to grow crops;

A_p represent the amount of land required to grow the food to feed one person; and

A_u represent the amount of land subtracted from agricultural use per person

TABLE 3   Land required to feed one person

Assumed type of agriculture	Kilograms of food produced per hectare	Kilocalories (Calories) produced per hectare	A_p, number of hectares (= 2.47 acres) required to grow food for one person
1963 U.S. Agriculture Providing Typical U.S. Diet[a]			.2
1963 Indian Agriculture—Only Cereal Grains Eaten	820	2,950,000	.30
1963 U.S. Agriculture—Only Cereal Grains Eaten	2,600	9,360,000	.094[b]
1963 Japanese Agriculture—Only Cereal Grains Eaten	5,480	19,730,000	.045

[a]The typical diet includes, for one person for one year, 42 kg. pig and poultry meat, 42 kg. beef and mutton, 250 kg. milk, 18 kg. eggs, plus cereals, sugar, etc.
[b]Assuming that 9.36 million kilocalories per hectare per year are produced, and that the average person will require .87819 million kilocalories per year, the area required to support one person is calculated to be .87819/9.360 = .0938 hectares.

added to the population.

Now $N + X = (A_f - XA_u)/A_p$, and we can find X by rearranging, which gives us

$$X(A_p + A_u) = A_f - NA_p, \quad \text{or} \quad X = \frac{A_f - NA_p}{A_p + A_u}.$$

This equation can be used to compute the probable number of persons who can be supported in the world under reasonable assumptions about agricultural productivity and land use, and also to predict the effect on agricultural production of changes in agricultural practice or land use. In either case, X will represent the measure of the effect.

Table 3 brings together data on the land required to support one person under four sets of circumstances.[10] Assuming agricultural productivity approximately that of the United States in 1963, and a mixed diet such as prevailed in the U.S. at that time, it becomes clear that the amount of land required to supply each person, with the several components of that diet (beef, for example) required large areas per unit of food produced. Assuming a level of agricultural productivity equal to that of India in 1963, where everyone eats cereal grain, the per capita requirement is higher. Assuming the 1963 production levels for the U.S. and Japan, respectively, and assuming at the same time that only cereal grains are eaten, the requirement is considerably reduced. Although higher productivity reduces the agricultural land per person required, all other things being equal, a mixed diet means a greatly increased requirement in agricultural land per person. Thus, in advanced countries, the gains as a result of improved production are offset by the greater area of land required for a mixed diet that includes meat. The figures in the third column of Table 3 are based on an assumed yield from cereal grains of about 3,600 kilocalories (calories) per kilogram.

Meat may soon become so expensive that it will disappear from the diets of many people—as is already happening among the middle class as well as the poor, even in the United States. To obtain an estimate of the number of people that could be safely added to the present world population, these figures have been combined with those on agricultural land lost to urbanization (Table 4).

If land use throughout the world were to follow the same trends as in the United States, so as to provide a similar diet, or if agricultural productivity throughout the globe were to reach an average level no higher than in India, the allowable maximum to be added to the world

TABLE 4   Relation of allowable addition
to present world population, agricultural production,
and land lost per person added

	Allowable addition to world population (billions)	
A_u: Amount of Land Removed From Agricultural Use Per Person Added to Population (Acres) / A_p: Amount of Land Required to Grow Food For One Person (Acres)	.287	.185
1963 U.S. Agriculture Providing Typical U.S. Diet— $A_p = .5$	4.87	5.60
1963 Indian Agriculture—No Animal Food Eaten— $A_p = .73$	2.95	3.28
1963 U.S. Agriculture—No Animal Food Eaten— $A_p = .23$	9.31	11.60
1963 Japanese Agriculture—No Animal Food Eaten —$A_p = .11$	13.22	17.80

population would be within a range of from 2.95 billion
to 4.87 billion people. The most likely prospect is for a
gradual increase in the standard of living throughout
much of the world, such as has occurred in Japan and
northern Italy. This would mean a world-wide increase in
the pattern of urban sprawl now found in the United
States, and an upgrading in the average diet. Under such
circumstances, the maximum allowable addition to the
present world population would be about 5 billion. As has
been suggested, the difficulty with this scenario is the

large input of energy, in the form of farm machinery, fertilizer and other components of agricultural technology, that would be required.

The answer to the question, "How many people can the world support?" depends entirely on the standard of living we are prepared to accept, and on the amount of energy that is put into agriculture. The more there is of urban sprawl, the less land is available for agriculture, and the fewer people the world can support.

One hazard of continuing the requisite heavy technological input is that any failure in the supply of energy would promptly doom a part of the world's people to starvation. It is as though a farmer were to maintain a large herd of cattle by bringing in extra food every day. If the farmer were to die suddenly, and no one replaced him in continuing to deliver the extra food to the cattle, one by one some or all of them would die.

The hazards inherent in massive application of fertilizer and pesticide are that they pollute soil and water, which may ruin waterways for sport fishing, and gradually degrade the soil. The use of fertilizer is associated with heavy monoculture, in which large acreages are planted out to one or possibly two crops. Under this type of cultivation both pest populations and plant diseases can build up and spread rapidly throughout an entire area. The highly productive agriculture which allows for a buildup in world population density is thus bought at a price.

THE U.S. BALANCE OF TRADE AND THE PRICE OF FOOD

Several billions could still be added to the world's population and be fed, however, under certain conditions (Table 4). Why, then, has the price of food begun to rise so

TABLE 5   U.S. balance of trade

Item	All amounts in billions of dollars			
	1968	1969	1970	1971
Total Exports	$34.199	$37.462	$42.590	$43.497
Total Imports	33.226	36.043	39.952	45.602
Exports–Imports	.973	1.419	2.638	−2.105
Food and Live Animals:				
Exports	3.890	3.733	4.356	4.365
Imports	4.577	4.531	5.375	5.531
Exports–Imports	−.687	−.798	−1.019	−1.166
Machinery and Transportation Equipment:				
Exports	14.447	16.403	17.882	19.465
Imports	7.987	9.763	11.172	13.904
Exports–Imports	6.460	6.640	6.710	5.561
Mineral Fuels and Related Materials:				
Exports	1.050	1.130	1.595	1.497
Imports	2.527	2.794	3.075	3.715
Exports–Imports	−1.477	−1.664	−1.480	−2.218

rapidly before the absolute limit to the number the world could support is reached? The reason is to be found in the pattern of world trade that is developing (Table 5).[11]

First, the U.S. has experienced an abrupt shift from an excess of exports over imports to a net trade deficit. Between January 1, 1970, and January 1, 1971, the national balance of trade worsened in three major categories: food and live animals; machinery and transportation equipment; mineral fuels and related materials. Given this development, what is the United States to do? There is little prospect of improving the balance significantly in manufactured products, given the quality of the products offered by competitors. Nor can the balance with respect to mineral fuels be improved; on the contrary, the imbalance is likely to worsen from now on. Clearly, the only solution is a sharp increase in the production and export of food. Meanwhile, however, the sharp rise in the demand for U.S. food in relation to the productive capacity, has led to a large price increase.

The figures in Table 5 suggest a basis for anticipating the pressure on U.S. food prices in the next few years. Net fuel imports in 1971, for example, were valued at $2.218 billion, and had been rising over the three previous years, at an annual rate of 14.5 per cent per annum. According to the U.S. Department of the Interior, net fuel imports (as measured in British Thermal Units) increased by 24.5 per cent in 1972. Assuming that the dollar value increased at the same rate, net fuel imports in 1972 would then be worth $1.245 × $2.218 billion, or $2.76 billion. Assuming as a conservative minimum a yearly increase of 14.5 per cent in the dollar value of net fuel imports—the 1968 to 1971 average, as opposed to the 1971–72 rate of 24.5 per cent—and further assuming that a deteriorating balance of trade in manufacturing compels us to raise the foreign currency needed to pay for fuel imports by selling corn, soybeans and wheat to other countries, how long could we do so before we run out of acreage for growing these crops?

We shall assume that in 1970 dollars, wheat from now on will sell at $3.00 a bushel, soybeans at $7.00 a bushel, and corn at $1.70 a bushel. Further, we shall assume a production of wheat over the next few years at 34 bushels per acre, soybeans at 28 bushels an acre, and corn at 87 bushels an acre. With these crops we can thus obtain $102 an acre for wheat, $196 an acre for soybeans, and $148 an acre for corn. We may then go on to assume that soybeans could be grown on 10 million more acres than in 1971,

TABLE 6   Acreages required for crops
to pay for imported mineral fuels

Year	Estimated value of net mineral fuel imports (in billions of dollars)	Land that would have to be planted to yield equivalent in foreign trade (millions of acres)		
		Soybeans	Corn	Wheat
1975	$4.14	10	12	3.96
1976	4.74	10	12	9.84
1977	5.43	10	12	16.6
1978	6.22	10	12	24.3
1979	7.12	10	12	33.2
1980	8.15	10	12	43.3

corn on 12 million more acres than in 1971, and wheat on 38 million more acres. If we then estimate the value of mineral fuel to be imported for the next several years (Table 6), it becomes evident that even under the optimistic assumptions used as a basis, the proposed strategy would succeed only until 1979. If the cost of fuel should increase faster than the cost of food, however, or if the increase in fuel imports should exceed 14.5 per cent, the strategy would fail at an even earlier date. Clearly, if we continue to pay for fuel with food, food prices can be expected to rise very rapidly indeed in the near future.

The early 1970s found us locked into a strange situation. For the moment, prices of energy were being held down, on the grounds that a sharp rise would constitute "anti-consumerism." Hence the demand for fuel is not driven down by higher prices, but is more and more in excess of the domestic supply. Consequently, more and more fuel is being imported, with a resulting deterioration

TABLE 7   Price trends in scarce foods

year	U.S. Clam Catch (Millions of Pounds)	Value of Clam Catch Per Pound	U.S. Crab Catch (Millions of Pounds)	Value of Crab Catch Per Pound
1950	41	$.27	159	$.06
1960	50	.24	222	.08
1965	71	.24	335	.09
1970	99	.29	277	.14
1971	83	.37	276	.19

in the balance of trade. To deal with this problem, we are selling massive amounts of food on world markets, thus bringing about an increase in the price of food at home. We are thus paying higher prices for food rather than for gasoline and natural gas.

Much of the current discussion of food prices contains a misconception about the nature of exponential growth. What is not clearly understood is the difficulty for the U.S., now that it is locked into a world-wide pattern of supply and demand for food, of altering the trend toward a demand that continues to outstrip growth in the supply, causing prices to go on rising exponentially. In fact, this has been the trend for some time where seafood is concerned (Table 7).[12] It is reasonable to expect that prices of meat, bread and other foods will follow.

A COMPLICATION: THE EFFECT OF AIR POLLUTION ON CLIMATE AND ON FOOD PRODUCTION

There has been speculation about the bad weather that has hampered crops recently in several countries—

Argentina, the Soviet Union, China, India, Pakistan and Indonesia, among others. Very hot or cold, or wet or dry years seem to occur with increasing frequency. Areas that once experienced an average of three bad crop years in two decades now have three bad years in a decade, or out of eight years. Why? Some scientists suggest that the world has been cooling off since 1945 and that air pollution may be implicated as a causal factor.[13]

Occasionally, a very large volcanic eruption ejects pollutants into the atmosphere in such massive quantities that the incoming radiation from the sun is deflected in part for up to several years thereafter. (The length of time depends on the size and specific gravity of the pollutant particles and the altitude to which they rise.) The most startling recorded instance over the last three centuries followed the eruption of Tambora (in the Dutch East Indies) in April, 1815. Its effects can be traced from newspaper accounts and temperature records. Over the entire period from 1701 to 1950, the average annual temperature for central England was 9.2 degrees C. In 1816 the average dropped to 7.9, and several months during 1816 and 1817 set records for cold weather—for example, July, 1816, was the coldest July between 1700 and 1950.[14] Such figures in themselves hardly indicate the magnitude of the effect. A drop in the annual mean temperature of 1.3 degrees C., or 2.3 degrees F., slight though it may seem to be, does in fact mean that the growing season would have been shortened, and that the amount of precipitation would probably have been greater.

Accounts from an English periodical, Evans and Ruffy's *Farmers' Journal and Agricultural Adviser,* suggest the economic and social effects of the cold weather, and a possible foretaste of what chilling of the planet by pollution might do in the future:

We had fine mild weather until about the 20th, when it set in cold, with winds at East and North-East, with partial frosts; these together have greatly retarded the operations in Agriculture, and very many cannot purchase seed corn, so that thousands of acres will pass over untilled, and sales of farming stock, and other processes in law, drive many of this useful class in society into a state of despondency. The wheats, late sown . . . have been partially injured by the frosty mornings. . . . Sheep and lambs have suffered from the severity and variableness of the weather. . . . The doing away the Income Tax and the war duty on Malt will afford some relief, but are wholly insufficient in themselves to restore this country to its former state of happiness and prosperity. (April 18, 1816)

From about the 9th or 10th of this month, we have never had a day without rain more or less, sometimes two or three days of successive rain with thunder storms. The hay is very much injured; a considerable part of it must have laid on the ground upwards of a fortnight. . . . Wheat is looking as well as can be expected, considering the deficiency of plants in the ground, and those very weakly . . . but still far short of an average crop. (August 12, 1816)

Throughout the whole month the air has been extremely cold; there has not been more than two or three warm days, being at other times rather cloudy and dark, and the sun seldom seen. The Oats . . . on high situated ground . . . are the most backward and miserable crop ever seen . . . for the greater part of the Wheat, where the mildew did not strike, has been very much affected by the rust or canker in the head. . . . There have been many seizures for rent this month, and many a farmer brought to nothing, and we hear of very few gentlemen who are inclined to lower their farms as yet; it seems they are determined to see the end. (September 9, 1816)

From these quotations it is evident that the effects of a very large volcanic eruption can be world-wide, not merely regional. A letter printed in the same newspaper reports further:

Last year was an uncommon one, both in America and Europe: We had frosts in Pennsylvania every month the year through, a circumstance altogether without example. The crops were generally scant, the Indian Corn particularly bad, and frost bitten; the crops, in the fall and in the spring, greatly injured by a grub, called the cutworm. . . . (November 10, 1817)

The precise effect of widespread air pollution, which builds up gradually is more difficult to determine with certainty because of the many other variables in operation at the same time. The effect of an extremely large volcanic explosion, however, is to hurl an immense load of pollutants into the upper atmosphere in a very short time. Thus we have, in effect, a gigantic experiment in the form of a pulse or shock wave, whose effects can be traced. The eruption of April, 1815 had, as one would expect from the preceding, a measurable effect on the price of flour traded in the London commodities market. Between June, 1814, and June, 1817, the highest asking price for a sack of flour of the best quality rose from 65 shillings, to 120 shillings, it was not until the end of December, 1818, that the price returned to normal.[15] The lag of 26 months between the time of the eruption and the time when the price was at its peak (from April, 1815 to June, 1817) is due first to the passage of time before the pollutants in the atmosphere reach their maximum effect in depressing the temperature, and second to the carryover of surplus grain kept in storage. Such a time lag is characteristic of complex systems, and is one reason why cause and effect are often not connected in the minds of many people: by the time the effect is evident, the cause has been forgotten.

Could the gradual world-wide increase in air pollution become serious enough to cause an increase in food prices? To answer this question, we must consider the current rate of increase in concentrations of pollutants, and the likelihood that controls on a global scale will be enforced. If loading the atmosphere with particulates continues throughout the world as a result of currently projected use of fossil fuels, by around the year 2040 the permanent load will be equal to that produced by the great eruption of Tambora.[16]

Will concentrations of pollutants continue to build up at this rate? There is ample printed testimony as to the difficulty of enforcing the limits on emission required of automobile manufacturers by 1975 under the Clean Air Act. Nor does a casual reading of petroleum industry journals suggest any great interest in slowing down the rate of growth in the use of petroleum products. Further, the developing countries are not interested in pollution control: they want more development. Without a really massive, world-wide change of attitude, such as does not seem likely at present, the nineteenth-century English farm newspapers quoted here may well be read as a scenario for the future.

FOOD FROM THE OCEAN

Some people, confronted with the preceding facts, will turn their thoughts to the ocean as a possible solution to the world's food problem. But the ocean is hardly a panacea. Catches of most important ocean fish have been declining, not increasing, in recent years.[17] Further, recent careful studies of the potential productivity of the ocean indicate that present totals of fish caught throughout the world add up to anywhere from a fifth to half of the total that could ever be taken from the seas.[18] Given that oceanic foods even at present are only a very small proportion of the world's total human food intake, this isn't very promising.

Why are the oceans such a poor source of food? There are two major reasons. In the first place, much of the open ocean is an aquatic desert: nutrients are constantly being lost as they sink to great depths, where no sunlight is present to make the production of green plant material possible, and from which thermal upwelling is insufficient to bring these nutrients back to the surface. In the second place, the harvesting of oceans differs from the harvesting of crops on land. On land, we harvest either green plants or animals that eat them (pigs, cattle, sheep, poultry). We do not harvest lions, tigers, wolves, bobcats or other meat eaters. One reason for not harvesting predators is the great loss of energy entailed. In any area, the proportion of the sun's energy available as food becomes less as that energy is transferred to a higher level—from grass to cattle to lion. But although we don't eat lions, we do take and eat their marine equivalents, the large predatory fish such as salmon and tuna, whose food is not plankton—the ocean counterpart of grass, which is too much dispersed to be harvested economically—but other fish at the end of a series of animals that eat other animals, only the first of which actually feed on plankton itself. This is the most inefficient possible use of the solar energy that reaches the surface of the ocean. As a result, the maximum amount of available food energy per acre of ocean surface may be very small. Also, much marine food is so low in calories per gram of edible portion that a great deal must be eaten to give a human being the energy to work.

OVERVIEW: MALTHUS WAS RIGHT

Behind the explosive recent increase in food prices are two messages: that, given the current ability of world agriculture to provide food, the world now has too many people; and that the United States is using up energy at a rate in excess of our ability to pay for it and still maintain a stable economy.

One immediate step that can be taken in this situation is for all families to stop having children, thus bringing about a reduction in the number of mouths to be fed per breadwinner, and with it a cut in the total food bill per breadwinner. Other steps, which require concerted political action, must also be taken toward making more efficient use of energy: greater use of public transportation and better insulation of buildings, among other measures.

One tragic misapprehension about our present situation has been exposed. We have thought of other countries as having far more energy to sell us than they actually do, and they have thought of us as having far more food to sell them than we actually do. But the truth will all too soon be clear to everyone.

4

ENVIRONMENTAL POLLUTION

Pollution has become a highly controversial subject. Advertising campaigns, along with many politicians, give the impression that the problem has been recognized, is not serious, is being dealt with, and is consequently diminishing in importance. Environmentalists, on the other hand, argue that the problem is extremely serious, is not being dealt with effectively, and is increasing in importance. The public is confused because of conflicting claims. Polluters have labeled pollution control as an enemy of full employment (e.g., "If the government forces high emission control standards on our plant, overhead will make the plant uneconomic, forcing us to close and put the entire staff out of work"). However, the man in the street can see that there is some basis for alarm, simply by looking out of his window at the sky, or at the water around the cities where he lives.

That the public is confused should not be surprising. Whereas articles minimizing the effects of pollution continue to appear in mass-circulation newspapers and magazines, statistical demonstrations of the effects of pollution tend to be published in technical journals not accessible to the public. Rather than attempt a comprehensive survey of all forms of pollution, this chapter will deal in some detail with air pollution and its effect on human mortality rates, and in less detail with the effects of air pollution on the weather, with noise pollution, oceanic pollution and its effects on fish and shellfish, and the question of aesthetics. Some remedies will be suggested.

HOW AIR POLLUTION AFFECTS HEALTH

One way to determining whether pollution is having a serious effect on human health is to compare death rates for polluted and unpolluted areas. Many sophisticated studies have shown beyond any reasonable doubt that air pollution is having a significant impact on mortality rates.[1] The conclusions that follow are in agreement with the findings of such studies.[2]

To illustrate the method used, four southern California counties, lying at about the same latitude, have been selected for comparison. In two oceanside counties (Santa Barbara and San Luis Obispo), pollutants are constantly being diluted by sea winds, whereas in the two others (Riverside and San Bernardino) lying in the interior, prevailing winds from the west carry a blanket of pollutants from metropolitan Los Angeles. This contrast between the high and low concentrations of pollutants is evident to the eyes, nose and common sense, and the pattern of distribution, including the way the polluted air moves in response to winds, is obvious from the air. Such observations are corroborated in data published by the California Air Resources Board. The pollutant most dangerous to health is probably oxidant, a gaseous agent which is believed to affect the aging rate of cells in the body. Oxidant is produced largely by the exhaust from automobiles. "Oxidant" as used here refers to a group of chemicals which

TABLE 1   Oxidant readings, August, 1971

County	Average of maximum hourly oxidant concentrations	Pollution level
Riverside	.23	High
San Bernardino	.19	High
Santa Barbara	.05	Low
San Luis Obispo	.04	Low

cannot be measured separately in the field: the most important constituents are ozone and peroxyacyl nitrates. The relative toxicity of the air in the four counties under consideration, in August, 1971, a high pollution month, is suggested by the oxidant readings (Table 1).

The possible effect of pollution on death rates is suggested by the data for the four counties collated in Table 2.[3] Two hypotheses are implicit here: first, that pollution has a significant impact on lung diseases, and second, that the total effect of pollution on the death rate is much greater than the figures on lung diseases would indicate. Of the two reasons behind this second hypothesis, the first is the difficulty an attending physician may encounter in correctly specifying the true cause of death—so that, for example, a person whose death is in fact the result of air pollution may be certified to have died of heart failure, even though difficulty in breathing caused by lung damage, leading to mild but continuous gasping, is what had put a fatal strain on the heart. The second reason behind such a hypothesis is that the most lethal effect of oxidant may not be on the lungs, but rather on the rate at which cells age. If it has such an effect, an increased death rate for all diseases, not just simply those affecting the lungs, would be evident. Careful studies of death rates from all U.S. metropolitan areas appear to support this hypothesis.[4]

TABLE 2 Air pollution and death rates

| County | Year | Population | Number of deaths | | | | Death rates per 100,000 per year | |
			Obstructive respiratory diseases A	Malignant neoplasms of bronchus, trachea and lung B	Both groups of lung diseases (A + B)	All causes	A + B	All causes
Riverside	1965	403,200	110	90	200	3,768	50	934
	1968	434,300	155	155	310	4,198	71	967
San Bernardino	1965	627,100	141	141	282	4,941	45	788
	1968	673,000	163	171	334	5,309	50	789
Totals and averages		2,137,600			1,126	18,216	53	852
Santa Barbara	1965	245,500	36	49	85	1,672	35	681
	1968	256,100	40	73	113	1,861	44	727
San Luis Obispo	1965	94,900	28	23	51	949	54	1,000
	1968	97,700	33	34	67	968	69	991
Totals and averages		694,200			316	5,450	46	785

Statistics for each of the four counties in 1965 and 1968 have been compiled on population; on deaths due to obstructive respiratory disease (emphysema, bronchitis, etc.) and malignant neoplasms (cancers) of the bronchus, trachea and lung; on the sum of the deaths due to these two groups of diseases; on the total number of deaths due to all causes; and on the death rates per 100,000 for the two groups of lung disease and for all diseases. From these figures (Table 1) it appears that pollution (1) increases the death rate due to lung disease and (2) causes an even greater increase in the total death rate. Thus, the average death rate due to the two groups of lung diseases does not fully indicate the effect of pollution on mortality rates. The difference in death rates per 100,000 per year between the two pairs of counties is only 53–46 = 7 for diseases of the lungs, and 852–785 = 67 for all diseases. If the data in this table have been correctly interpreted, findings based on mortality from respiratory diseases alone underestimate the total impact of pollution on mortality by a factor of 67/7=10 times.

But is it correct to interpret the data at face value, as has been done here? Statisticians point out many possible sources of error in such analysis.[2] For example, it has been assumed for purposes of discussion that pollution is the only cause of differences in the death rates between counties. There are, however, many other factors that might account either in part or entirely for the inter-county differences in death rates: e.g., differences in age structure (more old people), a lower income level, or a higher incidence of stress, accidents or crime in San Bernardino or Riverside counties. Or the residents might tend to be unusually susceptible to air pollution and of having migrated there to escape from air pollution elsewhere. Or there might be some environmental condition in San Ber-

TABLE 3   Stratification by age and death rate
in polluted and unpolluted counties

Death rates per 100,000 population in 1970 (all causes)

County	45–49	50–54	55–59	Average, for counties and ages
Riverside	548	870	1,236	937
San Bernardino	560	744	1,662	
Santa Barbara	464	688	1,209	785
San Luis Obispo	474	723	1,153	
Average	512	756	1,315	

Percentage by which death rate departs from average for four counties

County	45–49	50–54	55–59	Average
Riverside	7.0	15.1	−6.0	5.4
San Bernardino	9.4	−1.6	26.4	11.4
Santa Barbara	−9.4	−9.0	−8.1	−8.8
San Luis Obispo	−7.4	−4.4	−12.4	−8.0

nardino and Riverside counties—a fine, dry soil that is irritating to the lungs, for example—that tends to make them less favorable than the seaside counties.

To show how a citizens' action group, for example, might counter such objections, data were obtained from the California State Department of Public Health to carry this four-county analysis to a further state of refinement. To deal with the objection that the differences in death rates between the two pairs of counties are due to a difference in age, the death rates have been re-analyzed separately for each age (Table 3).[5] When correction for age is made, the difference in mortality between the pairs of counties actually becomes greater, not less. The uncorrected death rate in the two polluted counties (Table 1)

TABLE 4   Death rates per 1,000 population, 1968

State (polluted)	Mass.	Penna.	New York	Ohio	Illinois	Average
Death rate per 1,000	10.9	11.2	10.7	9.7	10.3	10.6

State (unpolluted)	Utah	Nevada	Arizona	Alaska	Hawaii	Average
Death rate per 1,000	6.7	7.9	8.3	4.9	5.3	6.6

was only $(852-785)/785 = 8.5$ per cent higher than in the unpolluted counties. Corrected for age, the difference is $(937-785)/785 = 19.4$ per cent. The same method can be applied to other characteristics of the population in the two sets of circumstances, for a full scale analysis of the impact of pollution on mortality.[6]

Such analyses, conducted here and in other countries, indicate that efforts to control smog are amply justified.[7]

Tables 2 and 3 do not prove beyond a shadow of a doubt that the higher death rates in Riverside and San Bernardino counties are due to pollution. They do show that death rates there are higher and that pollution is strongly implicated until some other cause can be found.

How much could mortality rates be reduced by cleaning up the air? A clue is provided by comparing the total death rates for states in which a high percentage of the total population live in urban areas, where smog concentration is high, with the total death rates for states a high percentage of whose population live in areas with a very low concentration of smog (Table 4).[7] Even from this very crude analysis, it may be concluded that clearing the air of pollution could reduce the death rate by $(106-66)/106 = 38$ per cent. More careful analysis of data on a city-by-city basis, with comparisons of death rates between polluted and unpolluted parts of the same city, suggest that this is an underestimate. All available evidence from British as well as North American studies suggests that a total elimination of air pollution could reduce the death rate to half its present level.[8]

Thus, almost any expenditure to control air pollution in big cities would be justified. The present effect of air pollution on death rates is much greater than most people realize.

NOISE POLLUTION

Noise can be controlled by such measures as fines that rise rapidly as the magnitude of the pollution increases. The noise level in all cities is rising, and the areas where noise is loud enough to interfere with talking or with listening to a radio, television or telephone conversation are becoming more widespread. But one form of noise pollution, that occurring under takeoff and landing paths of large jet aircraft, is so serious as to dwarf all others by comparison. In extreme cases such as that of the subdivision between the Pacific Ocean and the Los Angeles International airport, areas have already become uninhabitable because of noise. The affected area in Los Angeles has grown steadily since the 1960s and will probably continue to spread, so that in increasing numbers the population of the Los Angeles basin will have their sleep interrupted, and will be unable to communicate vocally or to hear sound communication of any sort for up to three hours a day. Several schools have had difficulty in conducting classes because of such interruptions.

If legislation were passed imposing a fine on airlines proportional to the number of landings and takeoffs they made that were intolerably noisy, and this fine were passed on to the traveling public in the form of increased fares, clearly there would soon be intense competition among airlines to reduce the noise of jet engines.

The problems of pollution, like many others associated with our civilization, stem from the use of matter and energy at high rates so as to increase productivity and the income derived from labor, rather than to improve the quality of life. As this high rate of resource use per capita spreads throughout the world, and people in increasing numbers aspire to share our life style, the dimensions of the international pollution problem will grow awesome.

POLLUTION AND THE WEATHER

During the past thirty-three years, the world temperatures have already been dropping, and pollution appears to be the most likely explanation.[9] Unless some other factor intervenes to alter the present trend, within a few decades mankind collectively will have polluted the air to a degree that may seriously affect the global climate.

How do we know this to be true? The consumption of energy throughout the world has been doubling approximately every decade. Thus, by the year 2000 the use of energy will be eight times as great as in 1970, and by 2030 it will be sixty-four times as great—if growth continues at the present rate. With such a level of pollution, the effect on the world's climate would be equal to what followed the largest known volcanic eruption—that of Tambora in the Dutch East Indies, one aftermath of which would appear to have been the doubling of the price of flour on the London commodities market (Chapter 3). Long before pollution became great enough to have so great an impact on the climate, other forces, such as inflation, would almost certainly have brought the global increase in the use of energy under control.

The example does serve to indicate, however, the hazards that would accompany inadvertent tampering with the climate as a result of air pollution on a large scale.

OCEANIC POLLUTION AND
MARINE FOODSTOCKS

Evidence of the effect of pollution on marine life is particularly startling at Escambia Bay near Pensacola, Florida.[10] A picture published in many newspapers showed the entire surface of the bay covered with dead and dying fish. Although the angle from which the photograph was taken made it difficult to estimate how many dead fish there were, a minimum estimate would be a few million. A lieutenant in the Florida Marine Patrol who was present at the site placed the total kill, when the entire bay was covered with dead fish, at between 50 and 75 million. This number becomes the more significant when we read that kills involving millions of fish occurred nineteen times in 1971, quite aside from dozens of smaller kills. In addition, a million-dollar harvest of oysters was totally destroyed. Because of the destruction of these food resources, and because of the powerful stench produced by the dead organisms on the surface of the bay, real estate values in the area are declining. The recreational fishery is dead, and tourism built around other water sports has waned.

What caused the pollution? Almost certainly, it is the effluents from four factories: a Monsanto Chemical Company plant that makes nylon yarn, an American Cyanamid Company plant manufacturing acrylic fibers, an Escambia Chemical Company plant that produces plastics and chemicals, and the Gulf Power Company. The pollutants affecting the bay are nitrogen, phosphorus and potassium, which cause excessive growth of algae. The presence of these microorganisms in turn brings about chemical and physical changes in the water of the bay so that the fish cannot survive.

The situation in Escambia Bay is occurring more and more frequently throughout the entire world. Although such bays account for no more than a small part of all water areas, they are certainly important for two reasons. First, because they are shallow, and because they receive the outfall of nutrients carried by streams, bays are vastly more productive than most other ocean areas. Second, and even more important, many species of ocean fish spend some part of their life cycle in shallow coastal waters. Some species breed in the shallow coastal waters, and migrate as juveniles out to the open ocean to develop to the adult state. In other species the larvae are spawned and developed in the ocean, and come inshore as adults. For both, poisoned inshore waters mean a break in the chain, so that the fish affected cannot survive even though the open ocean as a whole is still relatively non-toxic. Thus, long before the entire volume of water in all the world's oceans has become incapable of supporting life, the poisoning of bays could render them largely sterile. Far from providing an inexhaustible source of food *after* all terrestrial areas have been polluted or otherwise exhausted, the oceans would probably be dead *before* the land.

POLLUTION AND AESTHETICS

As recently as 1940, a small boy who went exploring ponds and streams at the edge of almost any North American city could still find fish, frogs and other aquatic life in abundance. A quick exploration of such streams and ponds thirty years later would show them to be almost devoid of life, for the simple reason that they are toxic. The sky over the south-central Pacific at sunset or sunrise is a splendid reminder of the clean air we could have everywhere if we thought it was worth the trouble. But it must come at a price—in dollars and in a curtailed expenditure of matter and energy.

THE REMEDIES

In considering how the growing effects of environmental pollution on public health, fisheries, agriculture, soil, water, air and climate can be abated, it is necessary first of all to examine the underlying causes of pollution. All processes by which matter or energy is converted from one form to another are inefficient. It is impossible to design a generating plant that will convert every molecule of fuel into useful energy, without waste, or a manufacturing plant that will convert each molecule of input into a desired product, with no material left over. The product of this inherent inefficiency is in fact waste, which—unless it is recycled—is the source of pollution. The more we produce, the more energy we use, the more there is of waste, and thus pollution, as a side effect or by-product.

There are thus two approaches to pollution abatement. One is to minimize the polluting effects of output by gathering up waste and recycling it at the end of the process. The other is to concentrate on input, cutting back on the extent to which processes are used. Common sense will indicate unmistakably which is more efficient. If processes are made more efficient, the same functions can be performed with less expenditure of matter and energy. A variety of measures that are available will be described in the chapters that follow.

5

MARKET SATURATION AND SLOWDOWN

One indication of a malfunctioning economy is market saturation. This is a sign that more capital is being allocated to manufacturing a product, or providing a service, than is warranted by the demand for that product or service. Since in almost any economy there are shortages of at least some necessary goods or services, some other use could have been made of the capital wasted in saturating a market.

Most economists have given little attention to this phenomenon, perhaps because so much economic thinking is based on the assumption that an economy can go on growing forever, whereas market saturation is a characteristic of reduced or terminated growth.

We are all familiar with the notion that the economy is expected to grow next year, the year after that, and so on indefinitely. But can it in fact do so? What happens when every family has three or four cars, several color television sets, more books and magazines than its members can read, and more phonograph records than they can listen to? What happens to a business that has been expanding at a rapid rate annually for many years, whose managers order more plant and equipment on the assumption that its growth will continue indefinitely at past rates, when that growth stops?

This problem is important to all of us, because over-production of manufactured products means unnecessary depletion of resources, needless pollution, and a waste of capital and of human effort that could have been used elsewhere. Market saturation, though bad enough if it occurs in a few industries, does not represent economic disaster, since capital and workers will migrate to other industries once selling the goods or services in question has become virtually impossible. If many industries encountered saturation of the market simultaneously, however, the economic consequences would be very serious, because of the difficulty large numbers of workers would have in finding jobs in other industries.

Could such a situation ever arise? U.S. industry is now able to produce most kinds of manufactured products faster than they can be purchased. For some time, the industrial capacity of the country has been only about 75 per cent utilized. Excessive capacity has not been regarded as a problem until recently, for the reason that major wars at fairly regular intervals utilized enough plant capacity to prevent the saturation of peacetime markets. Nevertheless, the problem is one that plagues not merely particular corporations but whole sections of the economy. For corporations facing a saturated market, the future is uncertain unless they redirect their activities. Meanwhile, the formation of conglomerates can be regarded as a defensive reaction to the danger of saturating a single market.

THE PERILS OF OVERESTIMATING RATE OF GROWTH IN DEMAND

Not only society at large, but also the industry in question pays a steep price for overestimating the rate of growth

TABLE 1   Sales volume and growth in demand

Number of years	3 per cent	5 per cent	7 per cent	9 per cent	Difference in sales at 3 and 9 per cent
One	103	105	107	109	6
Two	106	110	114	119	13
Three	109	116	123	130	21
Four	113	122	131	141	28
Five	116	128	140	154	38
Six	119	134	150	168	49

in demand for a product or service. Before considering specific examples, it may be useful to set forth the problem in the abstract. Suppose you are the president of a corporation whose sales have been growing at a certain rate for several years, but for which the market is close to being saturated—with the prospect that sales will expand at a decreasing rate in the future. What is the penalty for having assumed that sales will continue to grow at a high rate, if in fact they fail to do so? That the magnitude of the error is greater the longer one must plan ahead, and the greater the discrepancy between expected and actual growth rates, is evident to anyone with common sense. That the discrepancy could in fact bankrupt any business becomes clear when the expected volume of sales for each $100 worth of current sales is computed at four different rates of growth in demand, compounded annually (Table 1).

One reason for such errors as the over-building of hotels and luxury condominium apartments in Honolulu during the early 1960s is to be found in the corporate environment of 1960 through 1968, a period of rapid business growth during which those executives who were most successful—and therefore rapidly promoted—were those who took a repeated gamble that high rates of growth would continue. That period, however, was historically unusual.* Now that growth rates have slowed down, the most successful executives will be those who are more cautious in their anticipations. An overriding consideration will be the financial penalty for permitting a discrepancy between anticipated growth rates and realized demand.

AN HISTORICAL EXAMPLE: CONSUMER ELECTRONICS

Just how imminent is widespread market saturation? One broad category in which this state has almost been reached is that of home appliances, notably consumer electronics—phonographs, radio and television receivers, tape recorders, and combinations of these in consoles, with the most striking effects on table, clock and portable radios (Table 2).[1] The decline in sales has continued for long enough to leave little doubt that this is a permanent trend, rather than the short-term effect of a temporary economic situation. It is sobering to reflect on the consequences if this were to occur in a very large number of industries simultaneously.

A CURRENT EXAMPLE: PUBLIC AIR TRANSPORTATION AND THE AEROSPACE INDUSTRY

In 1970 four of the largest airlines (United, Trans World, Pan Am and American) all lost money. During that same year, indeed, fourteen of the top fifty public transportation companies throughout the country operated at a

---

* *Executives who are big plungers have already been replaced in large numbers by U.S. corporations. The educational publishing industry, for example, has seen a huge turnover of executives between 1968 and 1973 as college enrollments have ceased to grow at the 1960's rates and sales have dropped.

TABLE 2   Market for consumer electronics

Year	Sales of TV receiver tubes	Sales of radio and TV sets, record players, etc.
1955	$205,000,000	$1,508,000,000
1960	213,000,000	1,223,000,000
1965	432,000,000	2,463,000,000
1967	757,000,000	3,011,000,000
1968	738,000,000	3,084,000,000
1969	614,000,000	2,616,000,000
1970	458,000,000	2,168,000,000

loss. Linked with this saturation of the market for airline travel are the aerospace manufacturers. The seriousness of the situation for the entire economy becomes clear when it is pointed out that in size Boeing, Lockheed and McDonnell Douglas respectively rank seventeenth, thirty-third and forty-fourth among U.S. manufacturing corporations, and that their combined sales in 1970 totaled $8.31 billion—as compared with $8.73 billion for General Electric, the fourth largest U.S. manufacturer, and $7.5 billion for IBM, the fifth largest.

Thus, in terms of both earnings and number of employees, airlines and the manufacturers who supply them with aircraft together represent a large component of the U.S. economy. In 1970, 690,000 workers—about eight-tenths of one per cent of the entire U.S. civilian labor force, which then totaled 82,715,000[2]—were employed in making aircraft and parts.[3] If as many as half of these workers were to lose their jobs, 400,000 other jobs would probably be lost in an economic "ripple effect," thereby causing unemployment rates to become a political issue.

The aircraft industry is also extremely important to the U.S. balance of trade. In 1970, civilian aircraft, and parts for all aircraft, accounted for about 5 per cent of all U.S.

goods sold abroad.[4] Thus, the ultimate consequences of saturation in the world aircraft market might be a further drop in the value of the U.S. dollar in relation to other currencies. For this reason—especially since the statistics available are more complete than for any other industry—it is useful to examine the plight of the airlines and the aircraft manufacturers in some detail.

One myth should be disposed of at the outset. Airline executives have tended to blame the great excess of scheduled capacity over demand for seats, which led to serious losses in 1970, on the recession of that year. In fact, the recession only intensified a problem that had been developing for some time. As measured in revenue passenger seat miles, airline business had indeed been growing very rapidly for decades. But for three years prior to 1970, an ominous tendency to build up capacity (available seat miles) faster than demand (revenue passenger seat miles) had already been present, leading to a rapid decline in the load factor, or proportion of seats filled. A continuation of this trend means trouble for an airline, since it will generally lose money when the load factor drops below 47 per cent.[5] It is clear from a study of recent figures (Table 3) that the disastrous year of 1970 was the culmination of a trend that had been developing over the previous three years.[6]

TABLE 3   Overcapacity in airline travel

Year	Seat miles available (in millions)	Revenue passenger miles (in millions)	Passenger load factor
1966	137,845	79,889	58.0%
1967	174,819	98,747	56.5
1968	216,446	113,958	52.6
1969	250,846	125,414	50.0
1970	264,904	131,719	49.7

Figures for all U.S. airlines show graphically what happened: in 1966, a profit of $428 million; in 1967, a profit of $415 million; in 1968, a profit of $216 million; in 1969, a profit of $55 million; and in 1970, a $175 million loss.[7]

One might suppose that, confronted with this situation, the airlines would have reduced their purchases of equipment sharply. But that would have run counter to the interests of the aerospace manufacturers, who in the next few years were obligated to sell the airlines on the most expensive program of equipment replacement and fleet expansion in history, simply in order to survive.

The implications of this for the rest of the economy are so momentous as to call for spelling out in some detail. It should be borne in mind that purchases are being made for a market already evidently saturated with jet aircraft. Thus, the present conversion of U.S. airlines to the Boeing jumbo jet and the Lockheed and McDonnell Douglas airbuses represents one of the most gigantic and costly gambles in American corporate history—all based on the assumption that airline business would continue to grow at approximately 10 per cent per year until around 1980.[8] If this assumption is wrong, an enormous amount of capital will have been misallocated needlessly on a large fleet of white elephants which will only bankrupt their owners. However, not only can capital be wasted on planes, if growth projections are overestimates. There can also be massive unnecessary commitments for new airports, or airport expansion. Two examples are noteworthy. The Port Authority of New York has planned for a fourth major jetport for New York City at the former Stewart Air Force Base near Newburgh, New York, which has provoked a storm of controversy, because no one wants a jetport nearby. The opposition has been largely based on environmental impact. No one has thought to question the Port Authority's estimates of the growth of air travel. During April, May and June of 1973, there were serious discussions in Sacramento of enlarging the airport so that it could accommodate 500,000 takeoffs and landings a year, instead of 13,000, the present number. No one seems to have noticed that this involves an increase of 38.46 times, and that at the per annum growth rate in domestic revenue passenger miles that prevailed between 1969 and 1971 (1.73 per cent) it would take 211 years to get this increase in traffic. Further, with the current spectacular drop in the U.S. birth rate, the decline in the value of the dollar to about a quarter its value relative to gold in about 24 months, the drop in the stock market, and inflation, it is not clear that there will be any significant growth at all in air travel over the next two centuries. Few people seem to have grasped that the United States is now committed to a policy of importing fuel at a faster rate than it can export goods to pay for the fuel, and hence that the economic problem is endemic and chronic, as well as acute.

One central fact about new models of jet passenger aircraft has not been given sufficient attention. This is that, owing to the great costs involved in research and development, and in tooling up for assembly-line manufacturing of jumbo jets, airbuses or supersonic transports, a very large number of copies of each model must be sold before the manufacturer breaks even. Statements in print about what this number is are contradictory.

Daniel J. Haughton, chairman of the board at Lockheed, has said repeatedly at press conferences that his company needed to sell from 255 to 265 Tristar airbuses to break even.[9] A confidential Pentagon study of the Tristar program is reported to have shown, however, that from 300 to 350 of these planes would have to be sold before showing a profit.[10] Various other estimates of the break-even point on second-generation jets and supersonic transports put this figure within the range of 150 to 300 planes.[11] Very conservatively, each manufacturer of a new model plane in this class will have to sell somewhere between 180 and 350 copies to make a profit on the investment.

In that event, it must be concluded that the world-wide market for commercial transports is in danger of becoming totally saturated. As of October 6, 1972, the manufacturers of the Concorde, who would have had to sell about 150 planes to break even on their investment,[12] had received only fourteen firm orders—nine of them from airlines in France and Britain, the two countries sponsoring the project.[13] At $41 million a plane, the Concorde carries only 110 to 120 passengers, as against 330 or more for the jumbo jet, and makes a profit on each run only if 65 per cent of the seats are filled, as against a world average load factor of about 47 per cent.

As measured against any realistic assessment of future demand, it would appear that too large a number of aircraft models are being developed. To simplify the discussion of world-wide supply and demand, Boeing 707 "equivalents" is a convenient measure. Fully loaded, the 707 seats 137 persons. Expressed in these terms, the new models now being manufactured or in an advanced stage of development are as follows:[14] Boeing 747 jumbo jet, two and a half 707 equivalents (360 seats); Lockheed L-1011 Tristar, two 707 equivalents (272 seats); McDonnell Douglas DC-10, two 707 equivalents (272 seats); Japanese YX airbus, one and one-half Boeing equivalents (200 seats); Dassault Mercure, one 707 equivalent (140 seats); European Airbus A300B, two 707 equivalents (275 seats); British Airbus, one and one-half 707 equivalents (200 seats); Concorde, two 707 equivalents because of increased speed (110 seats).

Assuming that 300 copies of each of these planes must be sold to yield a profit satisfactory to the shareholders in the respective corporations, we have in prospect an addition to the world's jet fleet, aside from that of the Communist countries, of 4,350 Boeing 707 equivalents. As of March, 1972, a total of only 3,868 jet transports had been delivered to all airlines in the world, exclusive of the Communist bloc.[15] Thus, if manufacturers are to get a reasonable return on their investment, the entire fleet must be replaced or doubled. If the manufacturers' projections are correct, and passenger traffic grows 9 per cent every year from now on, a doubled fleet would be required in just eight years. This is unlikely, first because load factors in 1973 were very low—in fact, twice the actual load could be carried in the existing jet fleet—and second, because the compound increase of 9 per cent every year could be realized only if there were no repetition of the 1970 recession, during which passenger traffic, instead of increasing, actually declined by 2 per cent. In view of recent history, avoidance of another recession for eight consecutive years seems unlikely.*

That there is cause for concern is evident from figures showing a sharp drop in the number of jet transports being manufactured.[16] In 1967, manufacturers' shipments of such aircraft totaled 500; in 1968, the figure rose to 702; in 1969 it dropped again, to 509; in 1970 it dropped to 313, and in 1971 to 230. This decline is not due solely to the larger size of the aircraft being shipped. If every one of those shipped in 1971 had been a jumbo jet, 230 planes would still represent only 575 equivalents of the Boeing 707—a smaller number of jet transports than were shipped in 1968, and clearly a sign of saturation in the industry.

The aerospace industry would be in serious trouble if new orders for planes stopped coming in, since not enough orders for any of the new models have been received to

---

*Nothing has been said here about retirement of jet aircraft from the fleet because of aging. Much of the world jet fleet is in fact very new.

reach the break even point. The following figures give the situation, insofar as I was able to determine, on October 1, 1972: firm orders, plus options, contracts and prior deliveries for the Boeing 747 totaled 217; for the Concorde, 14; for the Douglas DC-10, 224; for the Lockheed L-1011, 184; and for the Airbus Industries A300-B, 31.[17]

THE ROLE OF GOVERNMENT IN MARKET SATURATION

The possibility that market saturation might lead to escalating unemployment had become serious enough by 1971 that the U.S. government decided to guarantee a loan by financial institutions to the Lockheed airbus project. Otherwise, the layoff of workers might have had political repercussions, along with the ripple effects in other industries. All the airbuses needed not only in the U.S. but throughout the world could have been produced by a single major manufacturer; but if only one, rather than the several that are now trying to slice up the pie, were in the business, there would already have been a calamitous spread of unemployment in the aerospace industry.

Government is in fact a party to the system that perpetuates market saturation. An important reason for over-investment in equipment by the airlines is to be found in the tax depreciation writeoff, which permits airlines to make money flying planes with a load factor of under 50 per cent. Another contributing factor is that the Civil Aeronautics Board grants rights to fly particular routes to so many carriers that none of them can make a decent profit. In a situation of this kind, the role of government is determined less by the public interest than by the wishes of special interest groups. Such a role is highly typical.

Government has tended, in fact, to work against free-enterprise capitalism, in which prices are determined by the free play of the market place, rising in response to increased demand and dropping as the demand decreases. Air fares are generally the result of interplay between airlines and the national or international agencies that regulate them. Icelandic Airlines, whose fee schedules for transatlantic flights are much lower than those of other companies, is not a member of the International Civil Aviation Organization. What if all international airlines operated free of the need to comply with international agreements?

In the airline industry, as in others regulated by the government, a fixed price structure does away with competitive prices, thus leading to competition over who has the latest models—all of which leads to waste and the excessive use of matter and energy. Price competition among airlines would go a long way toward limiting their present growth in the face of overcapacity. They can afford to run at only 50 per cent capacity simply because they operate under a cartel arrangement which sets a high price. If the cartel were broken, excess capacity would soon be cut.

WHAT ARE THE PROSPECTS?

In 1969, the last year for which complete statistics were available at the time of writing, sales of aircraft and parts by U.S. manufacturers were the largest they had ever been. The figure for non-military aircraft, engines, parts and all other aircraft products and services totaled about $12.8 billion,[18] of which only about $5.6 billion worth went into aircraft and parts. Of the latter amount, only a little over half went into the U.S. commercial fleet, the remainder having been sold to other "free world" countries. These figures suggest the enormity of the investment in a thousand airbuses, which would come to $17 billion—an amount that, wisely invested, could solve a number of major problems in the United States. The penalty for a wrong guess by the airlines could thus have national implications.

What is the probable outcome of this gamble? The annual review of airline traffic by the International Civil Aviation Organization for 1971 showed an increase in passenger miles of only 5 per cent over 1970.[19] Moreover, the growth rate in passenger miles appeared to be decreasing.[20] For U.S. airlines, the mean annual percentages of growth in domestic and international passenger seat miles since 1950 were as follows:

1950–55	1955–60	1960–65	1965–68	1968–70	1970–71
19.0%	9.8%	12.1%	18.4%	7.5%	3.1%

The hazards of predicting a high average growth rate in air transportation are evident: growth in this variable has a history of being very erratic. Moreover, projections of very high continued growth rates ignore the fact that whereas until recently some of the growth in air travel had been obtained at the expense of travel on trains, little further gain could be expected from that quarter. Passenger travel on U.S. railroads, calculated in millions of revenue passenger miles, declined from 31,790 in 1950 to 10,770 in 1970. During the same period, airline travel, calculated in millions of revenue passenger miles on domestic scheduled flights, rose from 8,007 to 104,156.[21] Although this gives a growth rate per annum of 13.7 per cent for the airlines alone, a percentage calculated from the totals for rail and air travel combined—39,797 million passenger miles for 1950, and 114,926 for 1970, for a growth rate of 5.4 per cent per annum—would be more realistic.

The prospects of future growth in the demand for air transportation can be assessed on the basis of air travel per capita since 1950. Table 4 gives the figures for revenue passenger miles on both domestic and international flights.[22]

TABLE 4   U.S. air travel

Year	Total passenger miles (in millions)	U.S. population (in millions)	Passenger miles per capita
1950	10,221	152.3	67
1960	38,873	180.7	215
1970	131,719	204.8	644

On the basis of these figures, a doubling of air travel per capita, as projected by the aerospace manufacturers for the next decade, would not appear unreasonable. A 9 per cent annual increase in traffic for the next ten years would mean that in 1980 the demand for seats would be 2.37 times that in 1970. Following this assumption, the "free world" jet fleet would require about the present number of planes, so long as the airlines remained willing to operate with very low load factors. Is this a reasonable expectation? Between 1940 and 1950, when commercial air transportation was just developing, growth rates in the demand for air transportation were of course much higher. But sooner or later the demand is satisfied, and its growth will then correspond only to increases in the population. Moreover, if the prospects for growth in the demand for air travel are based on projected incomes, a sharp per capita increase appears less likely. Described in terms of constant dollars, the per capita gross national product increased only from $3,522 to $3,597 from 1968 to the third quarter of 1971[23]—hardly enough to inspire confidence, especially when projected jet fuel prices are taken into consideration.

In a second possible scenario, the airlines would encounter much lower rates of growth in demand than expected, and would be obliged to cut back on orders for new planes so as to avoid going bankrupt themselves. This is not an unreasonable supposition. Pan Am announced losses of $34.6 million for the first half of 1972, as against losses of $39.5 million for the first half of 1971. United, American and Trans World Airlines would all be quite vulnerable to sharp losses in any economic recession, and would not be able to order new planes in quantity.

The economic repercussions of this scenario would be considerable. If the Lockheed Tristar program were terminated, for example, it has been estimated that 60,000 workers—about .07 per cent of the entire current U.S. labor force—would be laid off.[24] The "ripple effects" in the economy would probably lead to a much more serious situation, particularly since the Tristar labor force is concentrated in a small number of urban centers. Of the total number of workers who would be laid off if the project were canceled, 30,000 are in southern California. An increase in unemployment amounting to .2 per cent of the labor force would be politically explosive.

In a third possible scenario, the airlines, adhering to a belief in continued growth even in the face of evidence to the contrary, might go ahead and purchase all the planes that the aerospace manufacturers must sell in order to make a reasonable profit. When projected traffic demand failed to materialize, several of the airlines would go bankrupt, mergers would occur, and the financial position of the entire industry would become very shaky.

What conclusions can be drawn from all this concerning other industries, and concerning the economy as a whole? Clearly any belief that high rates of demand and of growth in demand will continue must carry a heavy penalty if that demand fails to materialize. On the other hand, faith in growth continues to be very strongly held. Even now, most people in the airline industry, and experts on the industry who serve as investment consultants, expect a traffic growth of from 9 to 11 per cent annually to continue for some time.[25] But what happens if the saturation point is reached for travel as it has for portable radios?

The number of cars manufactured in the United States will soon have reached a peak—approximately one car for every two persons in the country, not all of whom drive cars. The steel industry has supersaturated its market, both here and abroad. The overbuilding of resort hotels in Hawaii has already been noted; in the late fall of 1971 they were filling only 51 per cent of their rooms,[26] with 60–65 per cent the probable break-even point.[27] The railroads in 1972 had shown essentially no growth in business since 1966. The money being spent on education in many parts of the country had reached a peak, with the public balking at proposals to spend more, even though the number of students was rising.

In each of these instances, many experts would argue that the difficulties of the industry are purely temporary, brought on by the recession of 1970 and its lingering after-effects. By 1973, these experts insisted, the economy would once again be growing rapidly. But, why should an economy go on growing indefinitely at the same rapid rate? Nothing else does—not plants, animals (including human beings) or populations. Every system that has ever been observed grows rapidly during its juvenile stage, and after that at a decreasing rate until finally growth ceases altogether. Why should an economy be different? Common sense suggests further that markets can become saturated because at some point everyone has as much as he or she needs and can't be persuaded to buy any more. The limitation of time, if nothing else, puts a limit on what can be used. The very real limits on parking and driving space, and on human ingenuity, suggest that there will be a limit on the demand for cars.

A thoughtful reader will have noticed that the phenomenon of market saturation seems to apply only to certain components of the economy. Not many complaints have been heard recently about an excess of physicians, of clinical, health or hospital services, of housing for the poor in the ghettos, or high-speed rail transit systems, of symphony orchestras, or of small private liberal arts colleges.

We are not, then, faced with the problem of unemployment because of a shortage of jobs in which people could be gainfully employed. There is, rather, a shortage of jobs in manufacturing and other industries that use a small amount of labor relative to the consumption of matter and energy. A look at labor-intensive (i.e., service) industries will show many in which there is an acute shortage of labor. Consequently, the fundamental need is for a reordering of social priorities to increase the proportion of the population employed in service industries. Otherwise, unemployment will increase steadily from now on.

What would be the effect of widespread unemployment on those who are still employed? First, unemployed people use up capital that could be spent on other things if they were working. It would mean a decline in purchasing power, and less likelihood of their buying goods manufactured by those still employed. It would also tend to make workers fearful of losing their jobs, thereby causing them to purchase less than they normally would. Thus a vicious cycle would be set in motion.

Implicit in the saturation of markets are both pollution and the depletion of resources at a rate that could be avoided. A flight in an almost empty plane means a waste of fuel, plus wear and tear on the metal used in manufacturing the plane. Market saturation is thus only one symptom of an underlying economic malfunction.

6

INFLATION

Continually increasing population and an increasing demand per capita for resources are the root causes of a variety of economic and environmental problems. One of these problems is inflation, an exceptionally complex phenomenon attributable to at least ten distinct causes. Eight of those causes are ultimately traceable either to excessive use of resources or to an excess of population.

THE CAUSES OF INFLATION

Four causes of inflation stemming directly from excessive use of resources are:

1. Scarcity of a renewable (flow) resource. Wheat, beef and lobsters are examples.

2. Increase in the extractive cost of nonrenewable (stock) resources, after readily accessible deposits, pools or ore bodies have been exploited—a cause frequently aggravated by speculation. The fossil fuels and minerals, particularly precious metals, are examples.

3. Decrease in the buying power of a unit of money as a result of pollution control and environmental cleanup. The need for smog-control devices on cars is a familiar example.

Such devices would not be necessary if the population were half its present size, and there were only a quarter of the present number of cars per capita.

4. War. Because war always causes an enormous increase in the use of resources, and the demand consequently exceeds the supply, the consumer price index always rises spectacularly in wartime.

Two other causes of inflation, though not a direct result of excessive resource use, are related to it:

5. The artificial impact of advertising on demand. For example, car advertising encourages travel in private cars rather than by mass transit, which gives far more passenger miles of transportation per gallon of fuel and is much cheaper when all costs are included.

6. Manipulation of the money supply—i.e., the attempt to deal with social and economic problems by monetary rather than fiscal policy, which would mean a reallocation of the federal budget. An example of the latter would be to divert more government revenue to the construction of mass transit, and less to freeway construction. On the other hand, efforts to stimulate the economy by accelerating the rate of increase in the money supply, thereby speeding up the circulation of money, tend to encourage continued cost-push price increases.

7. A direct cause of inflation is over-population, which creates increasing per capita costs of maintaining social cohesion. A striking example is the cost per capita of maintaining police and fire departments in cities: the larger the city, the higher the cost per capita. What we have here, in other words, are diseconomies of scale.

8. Another cause of inflation is related not directly to population size but to the rate of increase. The faster a population increases, the more people of tax-consuming age there will be in the population as compared to those in

the tax-producing age bracket (Chapter 12). This leads to pressures for government services related to the young, thus adding to the impetus for government deficits, which contribute to inflation.

Inflation is also due to institutional and governmental defects:

9. Essentially oligopolistic bargaining between strong union leaders and strong industrial managers. When the organizations on both sides have a near monopoly in their respective fields, the prices of many products can rise above the competitive equilibrium. If society were more diverse, more options would be available, and competition would help to keep prices down, thus weakening the position of labor and management. Price regulation by government also allows prices of many products to stay above the competitive equilibrium. The airlines (Chapter 5) are an example.

10. In some cases, prices rise rapidly because the number of people to provide a service in great demand is inadequate. Physicians are a striking example. The available facilities for training physicians do not begin to meet the projected future need. If society were to invest less in using up matter and energy, more capital would be available, as a result of greater efficiency, to invest in training physicians. High doctors' fees are thus curiously related to the excessive use of resources. The same is true for many services: high wages in industry based on high throughput drive up wages and costs of services as a result of competition for employees to staff those services.

Clearly, if inflation is the result of over-population and an excessive use of resources, then it is the price we pay for our commitment to continual economic growth rather than to a stable society. The rest of this chapter will outline the prospects for the future, and suggest a remedy.

THE NATURE OF THE PROBLEM

The percentage of change in various price indices during March, 1973, adjusted to give an annual rate, will suggest the degree of continuing inflation: in the consumer price index (a measure of the value of money), 9.6 per cent; in groceries bought for home use, 37.2 per cent; in industrial materials other than food, 14.14 per cent; in consumer food prices, 30 per cent; in the prices of meat, poultry and fish, 72 per cent; in the price of services, 3.6 per cent; and in the price of food in restaurants, 10.8 per cent.[1] From these figures, it is evident that a salary-earner would need an increase of about 10 per cent a year just to stay even. For anyone on a fixed income, the situation is a catastrophe. The poor are hurt more than others because prices of food, the one thing they must have above all, are going up rapidly. At the same time, the price of labor was going up faster than its productivity, so that the real value of manufactured products per dollar was shrinking.

The situation becomes still more alarming when some of the ten causes of inflation are examined in detail.

The price of food is now rising around the world, simply because its entire population is large enough to place a serious strain on its ability to produce certain kinds of food. The huge, politically embarrassing food surpluses of a few years ago are no longer with us. Wheat is an illustration of what is happening. In 1969 it sold for $1.24 a bushel. In September, 1972 it was selling for $2.23 and in August, 1973 the price passed $4.00. In 1968, out of a crop of 1,577 million bushels, only 544 million bushels were exported. By 1972, 1.1 billion bushels out of a crop of 1.6 billion bushels were being exported.[2] As has already been made clear (Chapter 3), a continuous increase in the price of wheat may reasonably be expected.

In 1965, the price to the farmer per 100 pounds of beef was $19.90;[3] it had risen to $34.50 by September 19, 1972.[4] A startling example of how the human population can outdistance the supply of a resource is the northern lobster catch. In 1960, it was 31 million pounds for a landed value of $14 million. But by 1970, when its value had risen to $32 million, the catch was still only 33 million pounds![5] Despite the incentive of a price increase of over 90 per cent, a notable increase in the size of the catch had not been possible. One by one the same thing will happen to a great variety of renewable resources; unless the population stops growing there will be enormous price increases as demand overtakes supply.

One of the first minerals to run short in relation to demand is gold. At current rates of use, minable grades of gold-bearing ore will have run out, world-wide, by about 1988.[6] The result of this prospect has been a dramatic increase in the price, from about $36.40 an ounce in 1970 to about $126.00 in June, 1973.[7] To add to the pressure, as it becomes clear that gold and other such substances are running out, speculators will begin hoarding them in anticipation of a future rise in price.

The implications of speculation as a world-wide phenomenon are immense. Consider, for example, the position of the Middle Eastern countries and Venezuela with respect to crude oil. Suppose that whereas in 1972 they can sell a barrel of crude oil at the well for $3.30, by 1982 they can expect to sell it for $6.60. Doubling the price over a period of ten years means that to obtain the $6.60 they could receive in 1982, they would have to invest the $3.30 received in 1972 at 6.9 per cent interest, compounded annually for each of ten consecutive years—a rather high rate on a fairly safe investment. But suppose that by 1982, the price for crude oil should have risen to $10.00 a barrel at the well. In what could suppliers now invest their $3.30 so as to yield the equivalent of $10.00 in ten years? The answer is probably nothing! In other words, the smartest thing the supplier could do with his barrel of crude oil would be to leave it in the ground. It is easy to imagine what will happen to prices of all stock resources as it becomes clear that the world is rapidly running out of everything. Clearly, all the people of the world would then have a very strong motivation for cutting down the rate at which resources are being used up.

The more people there are, and the more gasoline they burn per capita, the more necessary control of air-polluting auto emissions will be. But this comes at a price. The total cost of all devices for pollution control, safety and reduction of damageability on full-sized U.S. cars will probably be at least $200.[8] In addition, the mileage per gallon of gas on the new car models will be lower.

The price paid for larger populations, and the increased institutional complexity required to service them, can be calculated with some precision. Table 1 shows the effects of population size in urban counties on three different components of the tax burden.

TABLE 1   Per capita costs of services

	Population category		
	Less than 10,000	50,000 to 99,999	250,000 or more
Police Protection	$5.70	$7.50	$19.10
Fire Protection	1.42	4.85	10.81
Sewerage	2.65	6.30	10.61

The picture is even more stark when we look at total tax expenditure per capita for local government. The latest available figures (mainly for 1967) are as follows: New York City, $892 per capita; counties with populations of 250,000 or more, $353; counties with 100,000 to 249,999, $275;

counties with 50,000 to 99,999, $246; counties with 25,000 to 49,999, $233.[10]

The argument is often used by developers and politicians that the public should approve of local development because it will "broaden the tax base and reduce assessments." The preceding figures show that this argument is spurious: as the population of urban areas goes up, so does the cost per capita for services, and so does inflation.

Bargaining between powerful unions and manufacturers, as a cause of inflation, can be best understood if we begin with a discussion of manufacturing firms and their economic character. One way to approach the subject is to imagine a set of manufacturing industries with ideal characteristics, and then compare it with what actually exists. To begin this exercise, a numbered list of characteristics can be drawn up so as to make a point-by-point comparison of the ideal with the actual situation.

The ideal manufacturing industry can be described under eight headings:

1. *Quality Of Product.* All products are made to last. Workmanship, raw materials and design are of high quality. Assembly-line procedures are designed so as to facilitate repair.

2. *Volume Of Production.* Since high quality is the aim, relatively small quantities of products are manufactured each year. This is the opposite of planned obsolescence.

3. *Competition.* The number of competing firms is large, so that the buying public has a choice of product and the price is set almost entirely by competition in a free market place. This is the current situation with respect to high-fidelity components, where there are a large number of excellent manufacturers, and the opposite of that in automobiles, aircraft and large computers—i.e., those retailing

for over a million dollars a system—in each of which the market is dominated by three or fewer corporations. In cars, the largest corporation does about 45 per cent of the business; in computers, its share is even higher.

4. *Number Of Workers.* Thanks to automation, the numbers are minimal. Assembly-line specialization, in which each worker performs over and over only one sharply circumscribed, simple task by rote, is avoided. Under automation, workers are assigned largely to the maintenance and monitoring of computer-governed robots. The workers no longer required for manufacturing are released for less routinized functions, particularly in service industries, research, development, invention, education and the arts.

5. *Prices.* Since they are largely determined by the balance between supply and demand, and by competition between manufacturers, the prices of manufactured products are kept reasonably low.

6. *Inflationary Tendencies.* Pressures in this direction are minimal, thanks to the balancing effect of competition, and the continual adjustment of supply to meet demand.

7. *General Consequences of This Situation.* Because annual rates of production are relatively low, it is unlikely that demand could suddenly be swamped by a glutting of the market with inferior products. Therefore, price changes are gradual rather than abrupt, a sudden drop in price never occurs, and there is little tendency toward over-expansion of plant and equipment.

8. *Implications In The Use Of Matter And Energy.* Under these ideal conditions, with no planned obsolescence, waste of matter and energy is minimized. Abandoned cars and appliances do not dot the landscape. Fuel is not wasted on engines that wear badly, and which lose efficiency but pollute the air increasingly with age. The loss of metals to rust

is minimized, since the rate of conversion of metals to manufactured products is cut in half. The depletion of resources is also minimized.

The present state of many industries differs remarkably from the preceding catalogue. The actual state of many industries may be described as follows:

1. *Quality.* A high incidence of defects, bad design, poor selection of raw materials; assembly technique not thought out so as to facilitate repair or ensure safety in operation. As a result, many items must be recalled by the manufacturer so as to eliminate hazards, and accident rates are high.

2. *Volume Of Production.* Since quality is low, the product short-lived, and the replacement rate high, the annual production is also high.

3. *Competition.* The number of competing manufacturers tends to be small, often bordering on monopoly.

4. *Number Of Workers.* Because of planned obsolescence, the numbers are unnecessarily large. With a high replacement rate, twice as much needs to be manufactured each year. Assembly-line automation is not pushed to the limit. In many industries, strong unions keep the numbers of workers unnaturally higher.

5. *Prices.* Since so many firms have the power of monopoly in the product market, prices tend to be high. Unions with monopoly power in the labor market bargain with a very small number of strong corporations for wage settlements that often have little or nothing to do with increased productivity per worker, or with increased demand for products. This is the "cost-push" effect: prices are pushed up by wage increases, rather than pulled up by demand—a process that has been repeatedly pointed out by J. K. Galbraith.[11]

6. *Inflationary Tendencies.* Since bargaining between unions and management can drive up prices even in the

face of a stable or dwindling demand, with little or no increase in worker productivity, purchasers get less for their money. When money declines, the result is inflation.

7. *General Consequences Of This Situation.* A time comes when the market is glutted with shoddy merchandise, and buyer resistance develops. When this occurs, prices decline, and corporations must cut back on the volume of manufacturing. But because wages have been fixed by bargaining with unions, the manufacturer cannot cut his costs by cutting wages. The only recourse is to cut back on the number of workers. This leads to a vicious cycle, since fewer employed workers means a reduced purchasing power—a situation that contributes to serious unemployment (Chapters 5 and 7).

8. *Implications In The Use Of Matter And Energy.* In many industries, matter is wasted as a result of planned obsolescence, and fuel is wasted as a result of decreased operating efficiency. The question of the effect on the use of matter and energy is not limited to that of policies *within a product line,* but also encompasses the broader question of which product line is most efficient—for example, the automobile as compared with technological innovations in mass transit consisting, say, of railways combined with a feeder system of minibuses.

What are the causes underlying the ideal and the actuality? Why is the present system defective in so many respects? The most important of these would appear to be planned obsolescence and the lack of meaningful competition, conjoined with the tendency of strong unions toward price-fixing.

A basic industry that illustrates these phenomena is steel manufacturing, which met its Armageddon in 1971. On August 1 of that year, a contract covering some 545,000 workers expired. In expectation of a long strike, the industry had built up a stockpile by manufacturing at a very high rate from February, 1971, until shortly before the contract expired. When a settlement was reached without a strike, the industry was left with an immense inventory, and production was cut back as a consequence. During the reporting period from August 7 to 13, 1971, for example, only 32 per cent as much steel was manufactured as during the period from May 8 to 14, 1971, and for the reporting period from December 18 to 24, 1971, the figure was still only 69 per cent of that in May. The effect on prices was soon evident.

In the Chicago area, where Number 1 grade industrial bundles of steel had sold at $39.10 a ton the year before, and at $31.10 a month before, the price dropped to $29.10 a ton.[12] This is compelling evidence that the wage demands of the unions had very little to do with demand for steel, but were the product of a bilateral monopoly—a very strong group of corporations bargaining with a very strong union, neither of which was influenced by possible competition from manufacturers abroad.

The ultimate consequences of this process are clear. Price rises that are not the result of increased value tend to debase the value of money and thus contribute to inflation. The triggering effects on wage demands by workers throughout the nation are obvious. But since price increases are linked by workers to hourly wages, any subsequent weakness in demand cannot be met by a reduction in prices but only by producing less—which means laying off workers. Thus union bargaining, oligopolistic product markets, market saturation and planned obsolescence, interacting as parts of an inexorable process, rush pellmell toward increasing unemployment.

What can be done about this insanity? Obviously, the one agency strong enough to fight the corporations and the equally powerful unions is the federal government. A number of mechanisms are open to it: first, it can increase competition by very strict enforcement of antitrust legislation; second, it can become much more vigorous in combating planned obsolescence, through careful government inspection and stiff penalties for selling defective or unsafe products.

But central to any realistic long-term assessment of the problem is the existence of an excessive labor force. A major contention of this book is that our society must shift its primary concern from manufacturing to services. One reason for this is an impending saturation of the market for many manufactured products. Tooling up to produce far more products than can be sold is all too easy for a modern industrial nation. To halt the process, the government must take the lead in hastening the conversion of a labor force oriented toward manufacturing to one oriented toward services and leisure activities.

If we persist in the puritanical view that this shift is in some way immoral, there can be no outcome but higher and higher unemployment. Between 1950 and 1970, the percentage of the non-agricultural labor force in the U.S. engaged in manufacturing dropped from 33.7 to 27.7 per cent,[13] with 16.2 per cent of non-agricultural workers in service industries and 18.1 in government.[14] This trend must be sharply accelerated, and since government is clearly the major employer of service workers, the acceleration will have to be accomplished through higher taxes.

As for the traditional notion that decency compels a certain limit on taxation, when the alternatives are to find a large proportion of the labor force on welfare—or, worse yet, starved and roaming the streets with violent and/or revolutionary intent—clearly the time has come for us to abandon that notion!

What must be understood about the problems related to manufacturing is that each of those problems constitutes part of a system that cannot be dealt with piecemeal. Inflation has arisen through the operation of several interlocking and interacting sets of causes. These consist of (1) bargaining between a small number of strong unions and a small number of strong corporations, plus (2) an excessively high proportion of the labor force in manufacturing, so that the outcome of the bargaining has a disproportionate effect on the entire economy, reinforced by (3) the virtual absence, because of the small number of competitors, of any meaningful price competition. Such a problem must be dealt with simultaneously on several different fronts: altering the mix of the labor force so that fewer people are engaged in manufacturing, an increase in competition, and exposure of prices to the free play of the market.

A principal reason for requiring government to deal with this problem systematically is the repeated demonstration by industry and the unions that neither one has any interest in looking at the system as a system, but that each is concerned only with its immediate self-interest.

To take one revealing example, on a single day in July, 1971, two different articles appeared in U.S. newspapers: one reporting that a number of banks had just raised their prime lending rate from 5½ to 6%, with the response of George Meany, president of the AFL-CIO, that "bankers should not be allowed to profit at the expense of the rest of the nation," and another, in an adjoining column, reporting that the United Steel Workers were expected to seek at least the 31 per cent wage boost which aluminum and can industry workers had gotten earlier in that year.[15] From this it would appear that organized labor thinks it is irresponsible for anyone to attempt to derive increased profit from the system—aside from organized labor. This failure to recognize that one's own group is just like all the others, however, probably is not confined to organized labor. The only possible conclusion to be drawn is that the federal government must be much more vigilant in protecting us all from one another and from ourselves.

PROSPECTS FOR THE FUTURE

Unless major social changes occur, we can reasonably expect that inflation will worsen at an accelerating rate. We are running out of everything; extractive costs for minerals and fossil fuels are going up; and sooner or later there may be speculation in resources. As the effects of pollution become more serious and are better recognized, the cost of pollution control will increase. So will the cost per capita of all social organization. Institutional and governmental defects and deficiencies can be expected to lead to great increases in the cost of services. Inflation will not be controlled, in short, without a heroic effort to deal with the multiplicity of causes that have brought it about. The rate of population growth must be slowed, both nationally and internationally. Ultimately, the world's population must decrease to a number that can be supported by the world's resources.

There must be a reduction in the rate per capita at which resources, both of matter and energy, are used. One way of doing so is to eliminate war. Stricter regulation of advertising will also be necessary. Greater reliance must be placed on fiscal policy, and less on monetary policy, in the solution of social problems.

A variety of measures must be taken to lessen the clout of unions. Here, as in manufacturing, the existence of a small number of very large organizations has led to massive instability in the absence of competitive pressure. If a single group has a monopoly on collecting the garbage in New York City, it will realize sooner or later that it can blackmail the city into paying almost any wages it chooses. If, on the other hand, a large number of private haulers were to offer competitive sealed bids for the contract to haul garbage out of New York, the price would drop, thanks to a tremendous competitive pressure toward being as efficient as possible.

An immense number of needed service occupations are at present badly understaffed, above all those concerned with medicine and health. It is a horrifying experience to visit the emergency receiving area of a large hospital late at night, and to watch an overworked staff try to deal with a volume of cases calling for ten times as many doctors. Education is similarly understaffed, particularly in such special areas as remedial reading, rehabilitation and specialized training for the handicapped. More research on arthritis, muscular dystrophy, cancer, and mental illness, among many others, is needed. So are out-patient facilities concerned with health. Far greater resources should be devoted to symphony orchestras, ballet companies, and other cultural enterprises, including noncommercial television. Specialized services could be made available to every social, cultural and ethnic group, to those plagued by alcoholism, drug addiction, marital difficulties, and suicidal or self-destructive tendencies.*

Clearly, strong action by government is needed. Why has that action not been taken? The answer, just as clearly, is that government is mainly responsive to large and powerful vested interests. Its failure to be responsive to the needs of ordinary citizens may be traced to the debts incurred by politicians at the time of an election campaign, and to the

---

*It should be noted, not simply in passing, that of the 55,000 people killed in the U.S. each year as a result of automobile accidents, one-quarter may in fact be suicides.[16] Senator Alan Cranston, in proposing an Improved Emergency Services Bill, estimated that 175,000 lives a year would be saved if the system of emergency medical care throughout the nation were improved.

lobbyists who devote themselves to ensuring that the will of their employers is carried out. To bring an end to this situation, taxes should be used to finance the campaign of any candidate who can collect a predetermined number of signatures on a nominating petition. As a result, no candidate could come into office with debts to be paid off to anyone. In addition, the activities of lobbyists should be restricted, monitored, recorded and published, and stiff penalties imposed on all violators. A new arm of government should be established to serve as an ombudsman and to counteract the effect on legislators of all other interest groups. The office of ombudsman should be sufficiently well funded and staffed to allow it to do research and planning on a large scale, as well as to do lobbying and to testify before committees on behalf of the public interest. The changes that must be put into effect if inflation is to be overcome are immense. But the penalty for not doing so is such as to justify the effort; for if the value of money were to drop at a rate of 10 per cent a year, no one can have any guarantee that his life's work will buy a secure old age, or that money put aside for emergencies will have its original value when it is needed.

7

A COMING GLUT OF MANPOWER

FLUCTUATIONS IN THE SUPPLY
OF NEW LABORERS

A phenomenon of immense importance that has received far too little attention is the variation in age cohort strength—i.e., in the numbers born each year.

A few figures will give some idea of how great the variation is. In 1935 there were 2,380,000 births in the U.S., and a net civilian immigration of 35,000. For 1957, the corresponding figures were 4,330,000 births and 272,000 immigrants; for 1969, 3,610,000 and 406,000 respectively.[1] Immigration figures have been included to show the relatively minor role played by new arrivals in population change throughout the nation. Earlier, the number of births in the United States had risen to a peak around 1915, tapered off and fallen sharply during the depression, the low point having been reached in 1933. A second peak occurred in 1957, and the number has been dropping since. In 1971 the total number of births was 2 per cent below the number for 1970, and it dropped still more in 1972. The number of births in any year will determine the number who, about twenty years later, try to enter the labor force for the first time. If the supply of those intending to enter the labor market is higher than the demand for their services, some of these young people will remain unemployed. Their decisions about marrying, having children, buying a home and what to purchase in the way of goods and services will be affected if the number of unemployed is high.

Thus, even though it might appear that a large number of young people reaching maturity would automatically ensure a rise in the demand for goods and services, without an increase in the demand for new workers the opposite may be true. In recent decades, the periods of greatest economic growth in the United States have in fact coincided with those when the number of young people trying to enter the labor force for the first time was small enough to be conveniently absorbed by the economy. Periods of economic hardship have coincided with those when an unusually large number of young people were trying to enter the labor force. For example, the depression years, from 1930 to 1940, came twenty years after a period during which the annual number of births had been very high. By 1953, twenty years after the annual birth rate was at its lowest, the economy was once more in full swing. It stayed quite healthy until 1969. Twenty years before, the annual number of births had been high. It would appear that if the generalization continues to hold, the economy may be quite shaky by 1977.

The difficulty here is in dealing with an extended time lag. Even were it to be clearly recognized by 1974 that the U.S. economy is in trouble, and that part of the cause is an oversupply of labor, what is there to be done about it, since the number of people trying to enter the labor force has already been determined? One thing we can do is to study the figures showing what to expect, year by year:[2]

TABLE 1   Size of labor force

Year	Number of births	Entry into labor force
1935	2,380,000	1955
1940	2,570,000	1960
1945	2,870,000	1965
1950	3,650,000	1970
1955	4,130,000	1975
1956	4,240,000	1976
1957	4,330,000	1977
1958	4,280,000	1978
1959	4,310,000	1979
1960	4,310,000	1980
1961	4,320,000	1981
1962	4,210,000	1982
1963	4,140,000	1983
1964	4,070,000	1984
1965	3,800,000	1985
1966	3,640,000	1986
1967	3,560,000	1987

These figures tell us a great deal about the future of the economy. If it does not grow fast enough to accommodate the young people who wish to enter the labor force, there will be *competitive pressure for jobs among the youngest workers for the next sixteen years.* We know that the year classes of 1950 and 1951 were accommodated into the labor force with some difficulty. Yet, as Table 1 shows, the year classes *every year for the next fourteen years* will be larger than that of 1950! Continuing to assume a twenty-year lag, only by 1986 will the prospective entrants to the labor force be fewer than they were in 1970. Simply to keep unemployment from exceeding the level of fall, 1971 (6.0 per cent of the prospective labor force), the growth of the economy must proceed at such a rate that for every 3.65 young people who permanently entered the labor force in 1970, no less than 4.33 can do so by 1977! This figure assumes that there will be no change in mortality rates during the first twenty years of life—as it is reasonably safe to do—and also that there will be no change in labor requirements as a result of further automation—an assumption that is far less likely. In other words, every year for the next fourteen years, the labor force must accommodate more new entrants than were *ever before accommodated in a single year of our history.*

What is the likelihood that this task can be accomplished? Examining the rate of growth of the employed component of the civilian labor force in recent years, we find that in 1960, 65.8 million people were employed in the civilian labor force; by 1970, the number had reached 78.6 million.[3] Thus, during a period of unprecedented affluence, the labor force increased by only 1.26 million each year.

It has already been argued here that no economy can be expected to grow forever at an accelerating rate, or even a constant one. On the contrary, it is more reasonable to expect that after it has been growing for some time, growth will occur at a *decreasing rate.* If this is true, the estimate that 1.26 new workers a year can be absorbed into the labor force is more likely to be high than low.

The only way that very large numbers of new workers can be absorbed into a stable economy each year is for an almost equal number of older workers to leave it through death, resignation, or retirement. Is it possible that a balance could be maintained this way over the next several years? No. Since the number of older workers in each age class who are approaching the age for retirement is less than half that of the prospective workers about to enter the labor force, the age structure of the population has forced us into a corner. On the one hand, the number of young people who will soon wish to enter the labor force is so large that a very high rate of economic growth will be needed to accommodate them; on the other hand, ours is a mature economy, in which it is hard to maintain a high rate of economic growth. If the situation were occurring in Japan, northern Italy, Greece, or Mexico, things would be different.

Many public leaders still insist that the number of children a couple desires to have is a totally private matter, to be decided by the couple alone. Yet twenty years later, when the children of all such couples are old enough to enter the labor force, their numbers will have become a very public matter indeed.

That long time lags in a system can lead to cycles such as we have been considering has been clear to economists for some time.[4] Some economists have also studied the relation of such matters as age structure and the number of people trying to enter the labor force to such swings in the economy.[5]

UNEMPLOYMENT AND BIRTH RATES IN THE YOUNG

If young people between the ages of 20 and 24 are indeed faced with economic hardship over the next fourteen years, what consequences are likely? In fact, the evidence available indicates that young people already see economic hardship looming ahead. According to the observation of the author, college students in the graduating class of 1973 were virtually unanimous on this point.

A Gallup poll to determine what percentage of the population regard four or more as an ideal number of children has been conducted eleven times since 1936. Since the results tend to correspond rather closely with data on the birth rate, this poll is an important indicator of child-bearing intentions. Between 1967 and 1971, it showed an astounding drop in the percentage responding that four or more children were ideal—from 40 to 23 per cent. More astounding still, among college students the drop over the same four-year period was from 34 to 14 per cent—a figure that was almost exactly the same for the entire sample between the ages of 21 and 29.[6] Among other changes is the discovery by the Census Bureau of a sharp drop in births to young women, and that whereas in 1960 about a third of all 21-year-old women were single, by 1970 the figure had risen to 50 per cent.[7] This tendency to delay marriage has possible implications not only for family size but also for the purchase of homes, among other investments. It is clear that economic pressure on young people to cut down such purchases could have a noticeable effect on the total volume of retail sales.

A phenomenon related to the economic squeeze on those between the ages of 19 and 24 is that it sets up a feedback loop that tends to intensify the troubles of those who reach the same age group a few years later. For example, elementary and high school teachers constitute 3 per cent of the U.S. labor force.[8] If there is a decline in births to those whose reproductive capacity is highest—those from 20 to 24[9]—the number of children entering kindergarten six years later, and likewise of students entering high school fourteen years later, will decrease. This in turn will mean fewer jobs as teachers for members of the same age bracket—and likewise fewer jobs for pediatricians, orthodontists, and manufacturers of toys, diapers and baby food, among others who provide goods or services for the young.

Exactly this phenomenon had begun to occur in the United States in 1973. Because of the reduced number of young people in public school, and the expectation of reduced numbers in high school and college, fewer teachers were being required, and young persons being trained as teachers in many states were wondering how they would earn a living.[10] Such young persons were in the same situation as prospective employees in steel, aerospace, automobile, or aluminum plants, or as prospective tomato pickers, university professors, airline pilots, electronics and railroad workers, tobacco growers, or toolmakers. In each of these fields, the labor force is already saturated, and the number of those wishing to join the labor force is greater than the number of positions open. In still plainer words, there are too many people and too few jobs. Moreover, the situation can only grow worse, since the number of those wanting jobs can be expected to increase every year until 1977, and to remain high until 1985. On the other hand, barring major social change, for various reasons that are to follow, the number of jobs is unlikely to increase at the same rate as the labor force.

Among the factors that can be expected to exacerbate the problem of unemployment is the fact that more women are entering the labor force. Another is that graduate school and military service are no longer effective in delaying entry into the labor force, in view of the scarcity of jobs for Ph.D's and the ending of the draft and the Vietnam war. Still another is the persistent refusal of labor unions to train adequate supplies of laborers—thus maintaining a scarcity of electricians, carpenters, etc.—and their resistance to admitting blacks and members of other minority groups. Union featherbedding, once an effective device for keeping up the demand for workers and minimizing unemployment, will be less significant as the importance of such industries as the railroads in American life declines.

The only possible way out of this situation is a change in the conception of what constitutes a legitimate job. A step in that direction is to alter the mix of functions within the economy—so that, for example, the reduced demand for teachers might be offset by an increased demand for staff at day-care centers. Eventually, however, the market even for such overdue services will be glutted. When that occurs, increased leisure, a reduced work week, and willingness to accept a stable or even a lower salary will all be necessary.

THE EFFECT OF "EFFICIENCY" ON THE DEMAND FOR LABOR

It may still be argued that however great the prospective increase in the supply of labor, the growth of the economy in the next decade will be rapid enough to absorb it. But before accepting the argument, it is necessary to inquire more precisely into where the increase in the numbers of jobs will come from. The government as a source of employment is uncertain, since the predictable reaction to economic slowdown at all levels of government is to put a freeze on hiring. Accordingly, it is to the private sector that we must look for the necessary expansion.

There is, however, a catch to the argument. Survival in the competitive world of big business has always meant a concern for "efficiency." One reason for this is that unions have made labor the greatest single cost factor: as we have seen, labor costs have risen faster than productivity, particularly in service industries. Thus "efficiency" is often synonymous with increased dependence on automation—which means replacing people with machines, and often a sharp curtailment of growth in the labor force. Unless there is a great change in the "mix" of the economy toward an emphasis on service occupations, the long-term implications are ominous.

Even a casual observation of many industries will suggest how pervasive the replacement of people by automation is becoming in our society. A commercial jet rarely employs more than three men in the cockpit (the pilot, co-pilot and flight engineer). As jets have increased in size, so as to accommodate 368 people instead of 80, there is clearly a decrease in the number of crew thus employed per passenger mile. As crops are sown and sprayed by airplane, and even fruit and vegetables are harvested by machine, the agricultural labor force goes on dwindling. Mining, the drilling of oil wells, and the operation of refineries tend increasingly to be under the supervision of a very small staff as more and more of the equipment used is controlled by computers. That banks, insurance companies, and other organizations that keep voluminous records are replacing people with computers is known to everyone.

A comparative examination of data on the number of workers required to deliver a unit of work over a period of years corroborates this impression. As the number of people employed per unit of added value diminishes, the quantity of materials and energy required goes up. In other words, as the productivity of labor increases, that of power decreases. This is a consequence of attaching too low a value to materials and energy, as compared to the value attached to human labor—and ultimately of certain assumptions fostered by the superabundance of resources that were once available to Americans (Chapter 11). The consequences of this value system become clear from a closer look at the data on airline operations, oil and gas production, and farming.

In airline operations, the work performed has been measured in revenue passenger miles flown on both domestic and international flights by U.S. scheduled air carriers. As Table 2 shows, there has been a startling increase in the "efficiency" of the airlines, with a concomitant de-

TABLE 2   Airline personnel, 1950–70

Year	Total personnel employed by scheduled air carriers (domestic and international)	Millions of revenue passenger miles flown (domestic and international)	Personnel employed per million revenue passenger miles
1950	82,786	10,221	8.1
1955	122,203	24,351	5.0
1960	162,771	38,873	4.2
1965	206,834	68,676	3.0
1968	295,025	113,959	2.6
1969	309,311	125,420	2.5
1970	290,700	131,710	2.2

crease in the number of persons employed per revenue passenger mile.[11]

The financial crisis that became acute in 1970 led to sharp cutbacks in personnel. As in most other industries, the reduction in business was perceived by the airlines as a signal to reduce staff, not increase it. Thus the problem of unemployment very quickly develops into a vicious spiral. Reduced business leads to layoffs, but since others are also laying off, there is a reduction in buying power—which once again means less business and further layoffs. Behind a general failure to recognize that this is essentially an uncontrolled feedback mechanism is the presence, during each of the post-1945 recessions before 1970, of a strong demand for growth in several major industries simultaneously, which could pull the economy out of a slump.

In need of more discussion are the implications for the entire economy when *all industries simultaneously perceive "efficiency" as the elimination of jobs, and convert to automation.* How, once this occurs, can we possibly absorb into the labor force the great year classes of young people we have been discussing?

The situation in the oil and gas industry further suggests how widespread the phenomenon has become. In 1950 these industries employed some 518,000 men to produce 18,290 trillion British Thermal Units of petroleum and natural gas. By 1960 the number of employees had dropped to 511,000,[12] while the production figure rose to 28,757 trillion B.T.U.'s; and by 1968 the number of employees was down to 467,000, even though production now stood at 40,680 trillion B.T.U.'s;[13] Stated another way, the number of employees per 10 quadrillion B.T.U.'s dropped from 284,000 in 1950 to 178,000 in 1960, and to 115,000 eight years later. Production has in fact more than doubled while the work force has actually declined. As has been suggested, the reason here is almost certainly an increased reliance on automatic systems of drilling for, transporting and refining gas and oil.

Farm workers have suffered a still larger displacement. Since 1950, about 5.3 million workers, amounting to more than half of the total work force, have been put out of work—while farm production as a whole increased by 41 per cent! From a total of 9,926,000 in 1950, the number of farm workers had dropped to 7,057,000 in 1960, and to 4,590,000 in 1969.[14] As the farm output index[15] rose from 86 in 1950 to 106 in 1960, and to 121 in 1969, the corresponding figures on labor per hundred units of output dropped from 11,000,000 to 6,700,000 to 3,800,000 in the corresponding years.

Given such effects of "efficiency" on employment, where are the jobs that will absorb the enormous numbers of young people who will be ready to enter the labor force in the next few years? Even normally optimistic business magazines are noting that employment in utilities, the steel and aluminum industries, automobile and tractor manufacturing, electronics and toolmaking has failed to rise. For prospective teachers at all levels, the employment situation is likewise bleak.

Assuming that income from jobs remains the basic means for distributing output, only two possibilities seem open for solving the problem of unemployment. The first is a major national commitment to dealing with large-scale social problems through massive construction and renovation.

Ultimately, the only cure for unemployment is a complete rethinking of the goals of society. So long as "efficiency" is a major goal, continually improved automation will go on displacing workers and leading to further unemployment. It is indicative of a defective and short-sighted system of values that "efficiency" is usually associated with the productivity of labor, rather than of energy and resources. We must substitute for "efficiency" the broader goal of finding productive occupations for the entire labor force. This will mean giving primary emphasis to services to human beings. Making a major commitment to this kind of activity is a way of dealing simultaneously with the problem of unemployment and with an acute labor shortage in many specialized services. If there were more street sweepers, gardeners, out-patient nurses, park maintenance and post office workers, the quality of life for everyone would be improved. If we continue to regard these functions as expendable whenever a minor economic recession occurs, then as the saturation of one market after another takes place, and as efficiency becomes the primary concern in all occupations, the problem of unemployment will become steadily worse.

HOW FAST MUST THE NUMBER
OF JOBS INCREASE?

The rate at which the number of jobs must increase so as to compensate for (1) an increasing labor force, and (2) increased productivity per worker, can be shown in a simple equation. If L represents the annual increase in the labor force, P represents the annual increase in the productivity per worker (including the effect of decreased hours per work week) and G represents the annual increase in gross national product in constant dollars, then the percentage of unemployment will remain constant only if $G = L \times P$. That is, the rate of increase per annum in the deflated gross national product will be sufficient to prevent rising unemployment only if the rate of increase is equal to the product of the annual rate of increase in the labor force, *multiplied by* the annual rate of increase in productivity per worker.

This is a tall order. The only way to achieve it is by a rapid shift in the mix of the gross national product toward the service industries. The problem can be dealt with on the side of demand as well as that of supply. At present, people don't buy more of the services they need because they can't afford them. If more money were put into the hands of such people, a shift in demand would lead to a shift in the employment mix. Thus, any policy to minimize the depletion and pollution of resources is necessarily intertwined with an income redistribution policy. (Income distribution policy is outside the scope of this book, but see reference 5, Chapter 15.)

There may still be those who will argue that unemployment can be cured without a major reordering of national priorities—that the economy as it was in 1973 can grow fast enough to absorb all prospective entrants into the labor force. Is there any possibility that they could be right?

This question can be answered most readily in terms of the preceding equation. The index of output per man-hour for the entire private sector of the economy, issued each year by the U.S. Bureau of Labor Statistics, rose between 1969 and 1971 from 107.5 to 109.9, for an annual increase of 1.11 per cent.[16] During the same period, the size of the labor force increased from 80,734,000 to 84,113,000, for an annual increase of 2.07 per cent.[17] When these two growth rates are multiplied together, we obtain a growth rate of 3.2 per cent per annum. In other words, this is the rate at which the nation's total productive capacity is increasing. If the gross national product in constant dollars does not increase by at least the same amount, unemployment will grow. To put the same thing still differently, if the total demand for goods and services does not increase at the same rate as the potential ability of the prospective labor force to provide those goods and services, some workers will be laid off.

As against this increase in productive capacity, the gross national product, as measured in 1958 dollars, rose from $724.7 billion in 1969 to $739.4 billion in 1971,[18] for an annual rate of increase of 1 per cent. Thus, the demand for production by the labor force did not equal the supply, and there was an increase in unemployment.

Some simple algebra will show us how the gross national product and unemployment are related.

Let E represent the proportion of the labor force that is employed. Then E will equal the number of employed workers in the labor force divided by the total number of employed and unemployed workers. The employment rate, expressed as a percentage, will then be $100\,E$, and the rate of unemployment, expressed as a percentage, will be

100(1 − E). If we wish to know how a change in the demand for labor as related to a change in productivity will affect the unemployment rate, we can let R represent the ratio of the new demand for labor relative to the new supply of labor productivity. Then R will equal the gross national product of the current year as related to that of the preceding year, both in deflated (constant) dollars, divided by this year's productivity as related to that of the preceding year. Accordingly, the new employment rate will be 100 ER, and the new unemployment rate will be 100(1 − ER). Now suppose we have an unemployment rate of 4 per cent. Using the figures just given, the new unemployment rate, projecting a 3.2 per cent annual growth in labor productivity and a 1 per cent growth of the gross national product in constant dollars, will be 100(1 − .96 × 1.010/1.032) = 6 per cent.

Given the present ratio of growth in the supply of and demand for labor, we can thus expect a rise in unemployment unless there is an adjustment somewhere else in the economic system. That adjustment could take the form of a shorter work week, government-stimulated demand for labor, or a massive government-financed program to keep young people in high school and college for a longer period, so that their entry into the labor force would occur at a later age.

What developments in the unemployment rate can we reasonably expect? As we have seen, there is every reason to suppose that for several years the supply of labor will be high in relation to the demand. We have also seen that productivity per worker will probably continue to rise at something like the rate obtaining in the early 1970s. It is less easy to determine whether at any time in the near future this growth productivity will have reached a plateau. A slowing down in the rate of growth in productivity per worker evidently did occur in 1969: whereas in 1968 the output per man-hour was calculated to be 1.033 times the output for the previous year, for 1969 the corresponding figure was 1.009 times that for 1968.[19]

When we consider the magnitude of the problem of market saturation that will soon be upon us with respect to the manufacturing of aircraft and aircraft engines, steel, automobiles, electronics, education, and elsewhere, the optimistic projections of those analysts who still insist that increasing production is no problem are hardly convincing. Perhaps the most important reason for the general absence of alarm on the part of economists is their failure to look at the several aspects of the problem as an interlocking whole. Although a number of elaborate computer-simulated models of the economy are now operating, none of them is yet complex enough to predict events more than a year or two away with any accuracy.

What we need are predictive models of the economy that include such phenomena as market saturation, age structure, the misallocation of capital, the effects of current labor-management relations on inflation, and the competition of foreign manufacturers. Only when such truly sophisticated computer models are in operation will it be possible to make informed decisions on how to manage the economy.

VARIATIONS IN UNEMPLOYMENT AMONG SOCIAL GROUPS

Politicians have frequently been guilty of two very serious errors concerning unemployment. First, they appear not to have perceived that although an unemployment rate of 6.1 means no immediate hardship for much of society, it does bring extreme hardship for members of certain categories. Among the most important of these categories are young people (almost 20 per cent of whom will be without work when 6.1 per cent of the labor force are unemployed) along with blacks and women. A further serious error on the part of politicians has been to downgrade the importance of very high unemployment rates among teen-agers. The one group of people who takes these high rates very seriously would appear to be the teen-agers themselves.

But today's teen-agers within a few years will be at an age during which birth rates are typically the highest. If, between the ages of 20 and 24, yesterday's teen-agers carry a vivid recollection of economic hardship, that recollection will be reflected in lower birth rates, since a high proportion of all births are to women under twenty-five.

The first businessmen to be aware of what politicians tend to ignore will thus be the manufacturers of diapers and baby food. Next it will be the toy manufacturers who discover that something unpleasant is happening—and then the politicians will be confronted with large numbers of angry eighteen-year-old voters. High unemployment among teen-agers is a condition that ought to be treated with respect, particularly among politicians who count manufacturers of goods for young people as their friends, or who are concerned with the teen-age vote.

Clearly it is important to study the probable trends of supply and demand for labor among young people in the next few years, since unemployment will have the most serious effect on the most junior members of the prospective labor force. In a population with an excess of those born between 1945 and 1965 over those born before and after, linked to a sharp decline in the birth rate for the next ten years, the long-term consequences will be historically unprecedented. By the year 2030, for example, a very large number of people from 65 to 85 years old will be relying for support on a relatively much smaller number under the age of 65—the ultimate legacy of a time when the size of one's family was thought to be a private matter!

It ought by now to be clear that no society can any longer look upon reproduction as a purely private matter without setting an economic time bomb for itself. Our present economic situation is the result of a collision between two major social processes: on the one hand, high birth rates beginning in 1945 have led to a rapidly increasing number of prospective entrants into the labor force; and on the other, a widespread concern with production and efficiency has prevented the number of new jobs from growing rapidly enough to absorb all those workers. Further, the recent high rates of growth have led to a distorted age distribution and a widespread social dysfunction where capital investment, taxation and social costs are concerned. In general, young people spend while older people provide capital and taxes. The social cost of young people to society is not only necessary but desirable; nevertheless, if it becomes intolerably large, growth of all kinds may be crushed. A self-regulatory feature is present, however, in a technologically advanced system, and it shows up in the sharply dropping birth rates that began to be noticeable around 1965.

The relation between birth rates and unemployment confronts us with a short-term problem of ameliorating economic hardship among young people who cannot find work, and a long-term problem of making sure we don't get into a mess like this again. Possible strategies for ameliorating the unemployment problem must begin with an assessment of how large the problem is. Having considered the *relative* contributions of different year classes to the labor force, we must then convert population statistics into *absolute* estimates of the number of prospective workers by examining the rates of participation by different demographic groups in the labor force. In general, we can expect to find that for any age, a lower proportion of women than men will be seeking to join the labor force. We can also expect to find that for either sex, the proportion of the population seeking participation in the labor force does not reach a peak until after the age of 24. The reason for this is that certain occupational groups require so long a period of training that the workers do not begin until they are in their late twenties. This is true of college professors and of medical specialists such as surgeons and psychiatrists.[20] The data summarized in Table 3 bear out our expectations.

TABLE 3   Participation in the labor force, 1970[21]

	Male	Female
16 and 17 years of age	46.7%	34.6%
18 and 19 years of age	68.8	53.4
20 to 24 years of age	85.1	57.5
25 to 34 years of age	95.0	44.8

For an accurate estimate of the number of people who will be seeking admission to the labor force for the first time each year, we should find out for each sex what proportion of the population first attempts in a given year to break into the labor force at ages 15, 16, 17, 18 and so on. Then for each year we should find out the total number of people in each of these demographic groups. For each combination of age and sex, we would then multiply the proportions by the total number in each group; then, by addition, we would obtain the total number of prospective new entrants to the labor force.

A simpler way to do this calculation, whose results are approximately correct, is to assume that the entire population first attempts to break into the labor force at age 20. Even though some people will have tried to break into the labor force at an earlier age, others will have done so at a later age, so that the two errors tend to cancel each other out. Using this method of calculation, the number of people attempting admission to the labor force for the first time in 1970 would be 2.55 million (Table 4).[21]

TABLE 4   New entrants to labor force, 1970

	Approximate population aged 20 in 1970	Participation in labor force at age 20	Likely number of new workers
Male	1.78 million	85.1%	1.52 million
Female	1.79 million	57.5%	1.03 million
Total			2.55 million

Because of death or retirement among older workers, the total increase in the labor force for the year will be somewhat smaller than this. Participation in the labor force by people in the ages just prior to retirement is very close to 60 per cent, and about 1.86 million people reach retirement age each year. Assuming that all retirements occur at exactly the age of 65, the total number retiring each year would come to about 1.12 million. Thus, the increase in the number of workers desiring admission to the U.S. labor force in 1970, stated approximately, would amount to 2,550,000 less 1,120,000, or about 1.43 million people.

Under current demographic conditions, then, the economy must provide 1.43 million new jobs each year, simply to keep the unemployment rate from exceeding 6.1 per cent of the labor force. How does this compare with the rate of increase in the number of jobs, and what would be the impact on the federal budget if this number of jobs were provided?

In fact, during all of 1971 the number of unemployed persons in the United States actually *increased* by 905,000.[22] One way of interpreting this figure is to say that the number of jobs being created by the economy increased only $(1.43 - .90)/1.43 = 37\%$ as fast as was required to keep unemployment at a fixed level.

This situation has led some to argue that the government should spend whatever is necessary to deal with the unemployment problem. How much would have to be spent, and could the nation afford it?

If the government were to hire 1,000,000 more workers at an average of $7,400 per person, the federal budget would have to be increased by $7.4 billion, or about 3 per cent more than the federal budget of $246.3 billion for fiscal 1973. The associated costs (extra office space, utilities, administrative overhead, etc.) would bring the total expenditure to about $14 billion—an amount that could be assumed without any increase if the government were to economize by cutting back on the purchase of manufactured products such as military and space hardware, and on the consumption of matter and energy (for example, by allowing less travel by government personnel).

This trade-off reveals in the starkest terms what from now on must be the alternatives behind all decision-making, public and private: is our objective to promote the consumption of matter and energy at the highest possible rate, or to provide the highest possible standard of living for the largest possible number of people? Clearly, very different fiscal strategies lie behind the two opposing goals.

Thus far, we have examined employment and unemployment in terms of the rate of increase in the numbers of the unemployed. Another aspect of this question is revealed if we examine the number of the employed, and the increase in those numbers. During the last six months of 1971, the number of people employed increased by 1.7 million, yet the unemployment rate did not decline.[23] The solution to this apparent paradox is that the U.S. population does not fall into two neatly separable groups, the employed and the unemployed. There are also the members of a third group, able to work but so discouraged about their prospects that they have not looked for work in over four months. The U.S. Bureau of Labor Statistics reported a total of 775,000 such people in the U.S. in 1971.[24] Thus it appears that when the economy does expand, the number of people unemployed does not drop, and may significantly increase. This happens because even though some of those who have been unemployed find work, their numbers among the identifiably unemployed are more than compensated for by additions from the third group that had previously been too discouraged even to look for work.

In other words, the statistics on unemployment published around the fifth day of each month do not tell the whole story. The complications in interpreting such statistics can be illustrated from the figures for December, 1971.[25] In that month the total work force consisted of 84.9 million persons (employed and unemployed), of whom 80.2 million held jobs and 4.7 million did not. The unemployment rate is not calculated from the ratio of 4.7/84.9, however, because the unemployment rate is expected to vary from month to month throughout a typical year. The government therefore corrects both the size of the work force, and the number of the unemployed, for this seasonal variation. Otherwise, the statistics for the twelve months would not be directly comparable with each other. After this seasonal adjustment, the unemployment rate is obtained as $5.2/(80.1 + 5.2) \times 100 = 6.1$ per cent of the labor force unemployed.

The catch in this statistic is that it ignores the approximately 800,000 unemployed persons who would be considered part of the labor force except for having become so discouraged about finding work that they have not looked for any in over four months. When we compute the unemployment rate for December, 1971, including this group in the numerator and denominator of the ratio, we obtain the new unemployment rate: $(5.2 + .8)/(80.1 + 5.2 + .8) = 6.0/86.1 = 7.0$ per cent of the labor force unemployed. This gives the basis for a more realistic assessment of what must be done to deal with the unemployment problem in the United States. A reduction of 1 per cent in the *true* unemployment rate would require that jobs be found for an additional 861,000 people. To get employment down to a politically acceptable 4 per cent of the true total labor force would require finding jobs for three times that many, or about 2.6 million people. Then, to keep unemployment down to that level, it would be necessary to provide about 1.4 million additional jobs each year.

What are the projects that could provide work on such a scale? No conceivable manufacturing projects could employ so many workers. Only about 60,000 are needed to turn out a new model aircraft, for example, and it has already been indicated that the market for aircraft is supersaturated. The total of those associated with the Atomic Energy Commission, including the staffs of contractors, comes to only 122,472 workers.[26] No scientific or developmental project could employ so many. The total number of scientists and engineers working in all of U.S. private industry is only 1.9 million,[27] and it would be difficult to invest, on short notice, in projects of sufficient scale to increase this number very greatly.

The difficulty is all the greater because of recent statements by heads of large corporations emphasizing that they cannot hire many more workers. Only about 74 per cent of U.S. industrial capacity is now being utilized, and it would be possible to increase production greatly without requiring much additional staff or equipment. In short, it would appear to be very difficult indeed to solve our present unemployment problem through major new manufacturing ventures alone.

Nor should this be surprising since, as has been pointed out, the whole thrust of industry has been toward minimizing the number of workers required to produce a unit of work, and since this goal has been successfully achieved. The consequence is that we now have a large increment in the population seeking employment, but that manufacturing is not available as a solution to the problem.

The one inescapable conclusion is that we can solve our unemployment problem only through investing in labor-intensive service industries. If we do not make this move, there will be a great increase in tax costs associated with welfare. Thus the choice is between spending money for greatly increased services, which will improve the quality of life, and spending a great deal of money on welfare programs, which return little to society.

INTERNATIONAL MONETARY TURBULENCE

The symptoms of international monetary turbulence are restrictions on trade, and sudden changes in the relative value of currencies (notably devaluation). Among the causes of such turbulence are several that originate in the disease of growth. Changes in the relative value of currencies are a reflection of differences in the rate at which money is paid out of and taken into a particular country. At least six groups of factors affect this rate. One is a difference in technological capacity. Thus, Switzerland exports watches and machine tools, Japan exports cameras, electronic equipment and ships, and the United States exports jet aircraft. Another factor is economies of scale. For example, only those countries with large numbers of highly trained engineers, scientists and skilled artisans, and with an enormous manufacturing plant and great capital reserves can compete internationally in the manufacture and export of tankers, computers and jet aircraft. Although the United States once led in the export of such products, other nations are now able to compete, thus eroding the U.S. trade position on such major products as automobiles (Table 1).[1]

TABLE 1   Trade in automobiles

Year	Exported (billions of dollars) value of road motor vehicles and parts	Imported (billions of dollars) value of road motor vehicles and parts	Surplus of exports over imports (billions of dollars)
1960	1.270	.627	.643
1965	1.744	.810	.934
1968	3.123	3.712	−.589
1970	3.245	5.068	−1.823
1971	3.873	6.847	−2.974

The United States still has a competitive position in the manufacture and sale of very large aircraft because so few nations or groups of nations have the manufacturing capability, the capital reserves, and the internal markets to sustain this kind of venture. For example, to make a reasonable return on capital invested in the jumbo jet and the airbuses, about 400 of any one model would have to be sold. The minimum population required to make use of that many such planes is about 70 million. A nation or group of nations with a smaller population simply would not have the guaranteed internal market.

A third factor is the demand for natural resources in relation to domestic supplies. For many years the United States was favored, but this situation has begun to deteriorate. The most striking example, of course, is crude oil (Table 2).[2]

This component of the trade deficit has increased by about 97 per cent in seven years, implying an average increase of about 10 per cent per annum. If we project this forward for another decade, the deficit then would be about 2.63 times the present deficit, or $7.47 billion! Since

TABLE 2   Trade in petroleum

Year	Value of petroleum products exported from U.S. (billions of dollars)	Value of petroleum products imported into U.S. (billions of dollars)	Trade deficit (in billions of dollars)
1964	.461	1.907	1.446
1966	.434	2.127	1.693
1968	.454	2.343	1.889
1970	.488	2.764	2.276
1971	.479	3.323	2.844

the United States will be increasingly dependent on imported supplies of fuel and minerals, our trade deficit will be increasingly sensitive to the efficiency with which we use these raw materials. The more wasteful we are, all other factors being equal, the more the value of our currency will decline in relation to that of other countries.

Tourism is yet another factor in the international movement of money. It had been hoped that devaluation of the U.S. dollar (making it worth 1/38 rather than 1/35 an ounce of gold) would discourage foreign travel by Americans. The International Economic Policy Association has noted, however, that many Americans were still choosing to incur the larger dollar costs of traveling abroad.[3]

The movement of international currency is affected by the availability of capital, either for purchase of goods or for investment. For example, India cannot accumulate huge foreign debts because she does not have the capital required to back them up, whereas the United States has a great deal of discretionary capital that can be invested abroad. Large investments of U.S. capital in other countries put many dollars in world money markets. This, in turn, is a major reason for pressure on the fixed dollar exchange rate, and for speculation against the dollar.

Finally, a very important cause of international monetary turbulence is the difference in the value of labor from one country to another. In Table 3, the 1971 figures are adjusted to reflect the currency alignment that occurred late in that year.[4]

TABLE 3
Average hourly compensation to factory workers
(in U.S. dollars)

	1960	1970	1971
Japan	$ .29	$1.06	$1.46
France	.80	1.67	2.01
Britain	.83	1.51	1.88
West Germany	.85	2.28	2.93
United States	2.64	4.20	4.46

Although by 1973 the price of labor was increasing elsewhere much faster than in the United States, our labor was still the most expensive in the world, and we could no longer compensate for that expensiveness with greater manufacturing know-how than that of potential competitors, or with low prices on inputs of matter and energy.

THE RESULTS OF INTERNATIONAL
MONETARY TURBULENCE

The immediate consequence of all these factors has been an adverse balance of trade for the United States. In a statistic called the "merchandise trade balance"—the sum of all exports, including military aid shipments, less the sum of all imports—the position of the United States has changed dramatically: whereas in 1970 the figure was never less than $71 million in any month, from April 1971 to September 1972 it was a minus quantity—ranging from −$212 to −$815 million—every month with the exception of September 1971.

This adverse balance triggered an important sequence of events. Since dollars were being spent by the U.S. in other countries faster than those other countries spent dollars here, there was an accumulation of dollars in those countries. In currency, as in everything else, the price is determined by supply and demand. Consequently, in those countries where dollars had accumulated, the price of the dollar tended to drop. If the drop was serious, the central banks of other countries could buy up dollars so as to reduce the supply in relation to the demand. But if the accumulation of dollars in other countries should become so great that the central banks of those countries could not shore up the price, there might be a different outcome. For example, on June 27, 1972, the Swiss government announced sweeping restrictions on the inflow of foreign capital.[5] These restrictions banned the purchase of Swiss securities and real estate by foreigners, and might remain in force for two years. At the time of the announcement, the dollar was worth 3.749 Swiss francs in unofficial trading—an improvement over the previous 3.715, but still below its legal floor of 3.7535 francs. Evidently the Swiss government had found that buying up dollars was not in its own interest, and that extensive dollars in Switzerland could undermine the Swiss economy, which is famous for its strength.

Another consequence of international monetary turbulence is devaluation. If a particular currency is overpriced, then the ultimate solution is to devalue that currency in relation to other currencies or to some absolute standard, such as gold. Such a move can have a number of deleterious effects. The dollar cost of overseas activities goes up, affecting not only normal international trade, but also investment, tourism and military activity. Thus, there is ample motive to avoid the necessity for devaluation by seeking to remedy its root causes.

THE INTERNATIONAL MONETARY SYSTEM

International monetary turbulence can develop when a government tries to speed up its economy through monetary rather than fiscal policy. For example, a central bank may lower its interest rates to encourage borrowing for investment in plant or business expansion. Money thus becomes more freely available, and this leads to a higher rate of inflation—which means that the dollar cost of U.S. goods to other countries is higher, so that they lose their competitive position internationally. Dollars accumulate elsewhere because more of them are being spent abroad than are being spent by other countries here—but also because when interest rates drop in the U.S., "footloose capital" migrates to other countries where the interest paid on investments is higher. These processes combined lead to a pile-up of dollars in other countries, and to a decline in the value of the dollar in relation to other currencies and to the free market price of gold.

The actions of the Federal Reserve Board can have an effect on monetary turbulence. The Board's low interest policy between December, 1971, when the currency was realigned, and March, 1972, for example, was unofficially estimated to have caused a capital outflow from the United States of more than three billion dollars.

Ultimately, the origin of international monetary turbulence can be traced to the high value the United States places on labor and the relative low value it places on energy and resources. But here another factor operates as well. The higher the proportion of our labor force engaged in manufacturing, the greater the effect on the balance of trade, since manufactured products tend to be exported whereas services are used domestically.

The relations between countries are affected not only by the per capita use of resources, but also by the size of the population. The greater the number of people, the greater the impact on the world's resources. This means that as the world population increases, competition between countries for the right to use various resources will intensify, finally developing into open conflict. As underdeveloped countries perceive that in selling their natural resources they are trading away their only chance for a decent future, they will begin charging higher prices. Ultimately, however, they may refuse to let us buy those resources at all (as in the recent Middle East crisis), backing up their refusal with as much force as they can muster.

Venezuela and the Middle East oil suppliers are already charging more for oil royalties. Some South American nations are becoming belligerent over fishing rights, declaring a 200-mile limit for their territorial waters, and threatening to seize and fine vessels from the United States, the Soviet Union, and other violators. The "cod war" off the coast of Iceland is another example. International conflict of this kind can only grow worse. Already, some American politicians are arguing that our fishing vessels should be given naval escorts to repulse any such attempts by a foreign power.

The intensification of such conflicts is taking place because there are limits to the earth's capacity to provide. This is true whether we are considering renewable resources, such as trees, whales and marine fish stocks, or non-renewable resources such as natural gas, crude oil, coal, uranium ore, chromium, nickel and tin. Renewable resources, if harvested too intensively, cannot grow fast enough to replace the stock at the same rate at which we are removing it. Non-renewable resources will simply be gone, and the faster we use them up, the sooner that will happen.

Critics who dismiss this as mere neo-Malthusian pessimism argue that human ingenuity can always come up with manufactured substitutes for depleted substances. What these critics fail to point out, however, is that the substitutes will be more expensive than the original resources; that they will intensify the problems of pollution as well as of depletion; and most serious of all, that human ingenuity is constrained by thermodynamic limitations. In plain words, we cannot solve all our problems with atomic energy. Nuclear power plants produce waste heat as a by-product, and the more such plants there are in a region, the more waste heat is generated.

It does not require much imagination to foresee the consequences if the heat thus generated became great enough to affect the polar icecap. Many of the world's greatest cities (New York, London, San Francisco, Rotterdam, Hamburg and Singapore, among others) are ports situated in convenient natural harbors. Any widespread use of nuclear reactors to provide the energy for extracting minerals from low-grade ores, for example, could quite conceivably melt the glaciers enough to raise the sea level, and flood these and other coastal cities.

Thus mankind is constrained not simply by a shortage of resources, but also by the laws of physics that govern the planet. There is nothing we can do about those laws.

All this adds up to the prospect of increasing tension between nations because there are too many people in relation to the available supply of resources. The per capita use of resources is rising rapidly in the underdeveloped world. There is no solution to this problem except by limiting the number of people in the world and at the same time decreasing the per capita use of resources in developed countries. The longer we delay going to work on a rational solution, the more serious our problems will have become. Our time is literally running out.

9

WRONG PRIORITIES,
MISALLOCATED RESOURCES

The amount of capital in the United States, though very large, is nevertheless limited. Consequently, if too much is invested in some projects, not enough will be available for other projects that may be needed. Projects in at least five broad categories, each of which calls for gigantic capital costs, are now urgent because of increasing population, resource depletion and pollution.

In the first of these categories are expenditures made necessary by increasing urbanization. The urban core must be renewed, if only because travel in cities will become too expensive if it must increasingly be routed around urban centers that have been abandoned. This implies a massive investment in renewal or replacement of housing. Mass transit will be required in order to relieve congestion, minimize pollution, and decrease the cost in energy of movement per passenger mile as fuel becomes scarcer and more expensive—and also because mass transit uses land more efficiently than cars. The organizational complexity of all branches of city government will increase, leading to an increase in the cost per recipient for services delivered (Chapters 6 and 12).

In the second category are those projects made increasingly costly by the scarcity of resources: the extraction of fuel and minerals from low-grade ores, and from deep or remote sites—e.g., those in the Arctic or under the ocean; the consequent need to transport resources for much greater distances; the investment in ocean-going vessels, gigantic seaports, pipelines, drilling rigs and refineries, including massive plants for the conversion of coal to gasoline.

Massive amounts of capital will also be needed for the development of entire technologies as conventional fuels run out, and a system powered largely by nuclear and solar energy.

The costs of pollution control will mount enormously as the urban population becomes denser, and as the necessary expenditure per person rises along with the absolute increase in population.

Finally, the present intercity transportation system, because of its inefficient use of energy, will have to be replaced as the present sources become scarce and expensive.

It is common sense to suppose that a project that is really needed by society will be undertaken. But although this was true in the past, such large numbers of astronomically expensive projects were never needed all at once—as they will be in the next three decades if the present system is to be maintained. The time may be coming when urgently needed projects will have to be deferred because too much capital has been wasted on other projects, which in retrospect will appear to have been unnecessary—wars, supersonic transports, new models of subsonic aircraft in excess, military and space installations built to gratify local pride, and many more. It is first necessary to put into perspective the very large amounts of money required for projects of such magnitude.

In the United States (and in France, but not England and Germany), a billion is defined as a thousand million and a trillion as a thousand billion, or 10^{12}. In 1972 the gross national product—that is, the total national output of goods and services valued at market prices—was about $1.150 trillion, or $1,150 billion in current dollars. This is the largest quantity to be mentioned in this chapter. Total federal expenditures in 1972 were in the neighborhood of $250 billion: a considerably smaller sum. Still smaller is the total capital investment in 1972 by all United States industries in new plant and equipment: about $92 billion. The total compensation of all employees in the United States in 1972 came to about $700 billion.

WHAT SOME NEEDED PROJECTS WILL COST

The $425 billion that could be required for nuclear power generating plants by the year 2000 has already been mentioned (Chapter 2). This amount would be almost five times the $92 billion invested by all U.S. industries in new plant and equipment in 1972—and although nuclear power would probably constitute the largest single requirement for capital (excluding only military preparedness), it is only one of a number of gigantic projects that can be foreseen.

The bill for urban mass transit can be estimated as follows: If new or renovated mass transit is required over the next two decades in each of twenty-seven metropolitan areas with one million inhabitants—a total population of about 60 million people plus an outlying population numbering an additional 20 million—and 50 miles of trackway for each million people is required,* the cost will be roughly $20 million a mile, for a total of $80 billion.[1] This is

*This estimate is based on the Bay Area Rapid Transit system (BART), whose 75-mile-long system will ultimately be expanded to a length of 200 miles and will serve four million people, including the population of Sacramento. This yields the rule of thumb: one million people require 50 miles of track.

not a large amount of money as compared to the $425 billion required for a new nuclear energy plant, but it is still large as compared to almost any other amount. (The estimated 1972 budget for the Department of Health, Education and Welfare was only $71.9 billion).

To clean up pollution, economists for the President's Council on Environmental Quality estimate that the combined cost to government and industry would be $316.5 billion dollars if zero discharge is the goal; to eliminate between 95 and 99 per cent of water pollution would cost only $118.8 billion. In a message to Congress on February 8, 1972, President Nixon urged an expenditure of $27 billion for the first four years of a water pollution control program.[2] Estimates in 1972 put the cost per car of keeping emission of pollutants to a satisfactory level at $150—which, assuming a sale of 10 million cars per year, would represent an annual expenditure of $1.5 billion.

Renewal of the urban core, because of the high cost of land, of demolishing decayed housing, and of constructing replacements, would mean a total capital cost of probably $10,000 per person housed. Assuming that 20 million people are involved, the total bill would amount to some $200 billion.

The probable cost of these large new projects (Table 1) is likely to come to about $52 billion per annum—or almost half the investment by all U.S. industry in new plant and equipment in 1972. To this amount must be added a lot of major expenses that have not been important hitherto: a tremendous increase in the number of railroad cars for shipping grain and fuel; an enormous investment in hospitals and the training of physicians to deal with an aging population; research on methods of generating energy, as well as on innovations in housing. Can we meet the bill? It is probable that we could, just barely—if too much capital is not allocated for gigantic "white elephants." The result of

TABLE 1   Prospective capital drains
on the U.S. economy

Project	Total cost over next 25 years	Cost per year
Nuclear Power Generation	$425 billion	$17.00 billion
Other Costs Related to Energy (Nuclear Fuel Enrichment, Coal Gasification, Tankship and Pipeline Construction, Ports, Refineries, Solar Energy Systems)	175 billion	7.00 billion
Mass Transit	80 billion	3.20 billion
Water Pollution Treatment	169 billion	6.75 billion
Air Pollution Control (Cars)	38 billion	1.50 billion
Air Pollution Control (Stationary Sources)	75 billion	3.00 billion
Urban Renewal	350 billion	14.00 billion
Total	$1,312 billion	$52.45 billion

the joint drain on the capital markets if the white elephants are undertaken along with the necessary developments would be dropping bond ratings and soaring interest rates. Because of those interest rates, borrowers, either in government or in industry, might be discouraged from undertaking socially mandatory measures.

MISALLOCATION OF CAPITAL

Apart from the military, two areas in which capital has been gravely misallocated in the United States are transportation and housing. The disparity between the five-hour plane trip across the country, and the trip of an hour or more from the airport to a downtown destination, is familiar to all travelers. Equally thought-provoking is the disparity between the half-empty plane and the frequent congestion on the ground. Could not the technical ingenuity that produced great speeds in the air be applied to these disparities? Even more dramatic is the contrast in Hawaii between the investment in hotels and luxury condominiums as compared to that in housing for the great mass of the population. In 1950, there were only 2,000 hotel rooms in all the Hawaiian islands. By 1960, the number had risen to 9,500. But by the end of 1972, it had risen to some 42,000[3]—about 8,000 of which had been completed in 1971 alone. As a consequence, by the early 1970s the rate of occupancy, which in Waikiki had stood at an average of 90 per cent or more, had begun to drop. In November, 1971, the average was 51 per cent.[4] Since the break-even point for hotels in Hawaii runs between 60 and 65 per cent,[5] it would appear that for many months in the future, more Hawaii hotels than not will be losing money.

Such wasteful expenditure of capital on building of hotels and luxury condominiums, in the face of an acute shortage of housing for the poorer sections of the population, is a phenomenon in many other cities. According to one recent study of the situation in Hawaii, the scarcity of housing was especially harmful to the elderly and those with low incomes; but since much of the housing supply was deteriorating, increasing numbers of the population were being affected. At least 40 per cent of Oahu residents were paying more than a fourth of their income for rent; the number of houses in bad condition had almost doubled in a decade, despite the demolition of nearly 1,000 units a year, until it was estimated that one out of every eight houses in the state was dilapidated.[6]

But the implications of misallocated capital also extend to the labor market. Where mass production and assembly-line techniques are employed on a large scale for airbuses, supersonic transports and space shots, a labor force is being developed for which there is no permanent need: these employees are faced with the prospect of future unemployment. If, on the other hand, the labor force is put to work on rebuilding cities and expanding mass transit, the demand for their services can be expected to continue. Similarly, a program of investment in solar energy collectors and fusion reactors, to replace gas, oil, oil shale and coal as these fossil fuels run out, must begin while capital is available— not after immense sums have been spent on projects that are not necessary.

THE CAUSES OF CAPITAL MISALLOCATION

Why do vast amounts of capital continue to be invested in such projects, while others that are desperately needed go begging? That projects such as the supersonic transport will stimulate economic growth, reduce unemployment, improve the balance of trade, and benefit stockholders are standard arguments. The U.S. supersonic transport and the British–French Concorde (Chapter 5) provide a classic example of what can happen. How, then, did the costly error of deciding to build the Concorde occur in the first place? And how did the United States come so close to investing $250 million in building two prototype supersonic transports that would have retailed at the even higher price of at least $80 million each? How could we have failed to detect the irony, at a time when the Japanese were devoting their own engineering skill and manpower to the development of high-speed trains? Now that all the arguments put forward in favor of investing in supersonic transports have turned out to be spurious, it is essential to understand what error in decision-making is responsible.

Many of the men charged with decision-making in government and large industries at the time had arrived at their positions during a period of unusually high economic growth. Between 1950 and 1969, for example, revenue passenger seat miles flown by U.S. scheduled airlines on domestic flights rose from 8 billion to 102.7 billion—a twelvefold increase over just nineteen years, for an average increase of 14 per cent annually.[7]

But why the failure to invest in needed projects? A case study that may shed some light on this question is the nuclear enrichment industry, which converts natural uranium into nuclear fuel. The capital costs involved in building enough plants and providing the power needed to produce enough uranium fuel for the year 2010 would be $25 billion.[8] Moreover, because of the problem of obsolescence (Chapter 2), reactors will not necessarily be economically operational. Thus, the assessment of risk becomes important in determining what options are preferred.

Equally important is that most senior decision-makers in government serve terms of two, four or six years, and that for corporate decision-makers in general the term is likewise short. As a result, the horizon of most planning is circumscribed by what offers a big payoff within a short time span, rather than what might be good over the long term. Thus the decision-maker has every incentive not to consider long-term consequences.

PUBLIC VERSUS PRIVATE SECTORS OF THE ECONOMY

As John Kenneth Galbraith[9] and others have pointed out, a fundamental issue concerning the use of capital is our relative unwillingness to spend on the public as opposed to the private sectors of the economy, simply because of our great commitment to free-enterprise capitalism. At the heart of the issue is taxation. Many people appear to regard taxes simply as money taken from them. But taxes have a number of functions besides payment for services. They provide large amounts, per dollar of taxation, for wages; and since a major social difficulty in the next few years will be unemployment (Chapter 7), those amounts can be expected to grow. As the victims of more than one revolution have learned, too late, there are reasons other than ethical or religious for sharing one's wealth with other people.

The public sector is so starved for funds that *Time* magazine for March 13, 1972, ran a cover story entitled "Is the U.S. Going Broke?" In 1960, voters throughout the nation approved 89 per cent of all school bond issues put to a vote; in 1965 the figure was 77 per cent, but by 1970 it had fallen to 48 per cent.[10]

What is most remarkable about the growing unwillingness of voters to pay taxes is the actual cost of inadequate taxation—fewer police, a narcotics trade almost out of control, and a consequent enormous petty crime rate, often associated with random assaults by addicts trying to steal enough for a fix. Similarly, at a time when welfare rolls are high, cutting government expenses by firing very large numbers of local employees means only that those expenses will reappear as additional payments to welfare recipients—who this time perform no work in exchange for the tax money they receive.

Unwillingness to invest in the public sector can be insanely short-sighted. In many states, facilities for training physicians are starved for funds, at a time when medical services are a major component in the increasing cost of living. Whoever heard of reducing the cost of an overpriced service by cutting down on the supply? Yet that is precisely what has happened.

Under-investment in taxes has the insidious side effect of contributing to the depletion of resources and to pollution, both of which are endemic to the private sector. An increased expenditure of public funds would lead to greater efficiency in the use of matter and energy. Increasing the investment in public transportation, for example, would almost certainly entail the development of railroads, more and more of which are publicly owned and operated, which make vastly more efficient use of matter, energy and land, and which are responsible for less pollution per passenger mile than other modes of transportation.

Institutionalized government planning such as Rexford Tugwell has proposed[11] would, if adequately staffed, greatly broaden the horizon of public decision-making. Since projects such as, for example, those associated with nuclear power, involve gigantic risks, government must offer financial incentives if corporations are to undertake them. In corporations, planning must take cognizance of the hazards of assuming too high a future growth rate as against the possible rewards.

COMMITMENT TO ECONOMIC GROWTH

Growth in the past served to improve the standard of living, and was an incentive to those entrepreneurs who helped others as they advanced themselves. Because of these historical advantages, the thinking of many individuals and the policies of many institutions have become ingrained with a commitment to continuous growth. We have now reached a point, however, at which the deleterious consequences of growth outweigh its benefits. Thus, we need to commit ourselves to a redistribution of the fruits of growth and to stable economy. But even though many variables show clearly that growth is tapering off, there is a tendency on the part of institutions, including the government, to give the impression that growth is still occurring. The effort to do so is in itself harmful, and many people will suffer as a result of the assumption that past rates of growth will continue indefinitely.

SOURCE OF THE COMMITMENT TO GROWTH

The roots of this commitment to growth go deep. An instinctive preference for expansion shows itself in the sadness nearly everyone feels on witnessing the death of a resort, as fewer and fewer of its summer guests return, or a town as the last industry pulls out, or on discovering a ghost town in mining country or the desert. The arrival of new people means a more stimulating, diverse and rewarding life, emotionally as well as economically. Greater density of population, up to a point, means increased economic rewards for everyone. Comparative figures for 1970 on average hourly earnings of industrial workers in four states with low density and little urbanization, and in four others with high density and large metropolitan areas, show the correlation of higher wages with population density and urbanization (Table 1).

TABLE 1   Wages and population density, 1970

State	Average hourly earnings	Population per square mile	Metropolitan population as percentage of total
Maine	$2.71	32.1	21.6%
Mississippi	2.43	46.9	17.7
Arkansas	2.48	37.0	30.9
North Dakota	2.93	8.9	11.9
Michigan	4.15	156.2	76.7
California	3.80	127.6	92.7
Washington	4.06	51.2	66.0
New York	.3.46	380.3	86.5

The rise in wages with rising population appears, however, to have come to an end in the United States (Table 2).

Clearly, given the time lag in most human perceptions, the majority of the population cannot be expected to realize immediately that the benefits accruing from continued economic growth are at an end. Even stronger will be the resistance on the part of those persons and institutions who have

TABLE 2   Real Wages, 1950–71

Average gross weekly earnings less social security and federal income taxes, for workers in manufacturing with three dependents (in 1967 dollars)

1950	1955	1960	1965	1970	1971 (April)
$78.17	$87.02	$90.32	$102.41	$99.66	$101.11

been most successful in garnering the benefits of growth: the astute land speculator, the shrewd investor in common stock, or the corporation which through brilliant management has exploited a rapidly growing market for a new product or service. That such benefits cannot be exploited indefinitely, however, will be clear from a study of the motor vehicle industry, which has completed its growth. During the first two decades of this century, motor vehicle registrations grew from around 8,000 in 1900 to just under 500,000 by 1910, and by 1920 to around 9,000,000. Since 1930, however, the rate of growth has been much slower. Registration went up in that year from 27,000,000 to 30,000,000 a decade later, to just under 50,000,000 in 1950, to 70,000,000 in 1960, and to around 100,000,000 in 1970.[1] Sales of tape recorders (including imports) have grown since 1960; once again, however, the rate of increase has become less steep: after rising from 300,000 to 3,400,000 by 1965, and to 8,400,000 by 1970, with an increase of nearly a million and a half in the single year from 1969 to 1970, the annual total in 1971 went up by barely a quarter of a million.[2]

The rise in profits while a corporation or industry is in the initial phase of growth naturally sets up an enormous pressure to sustain its growth at the same high rate. Since salesmen and the marketing branches of corporations can profit by the interest of potential purchasers, they have a motive to maintain growth. Corporations are eager to report higher rates of increase in earnings to their share-holders, and the manager whose income depends on the appearance of corporate or government health has a vested interest in contributing to that appearance.

Growth is also of great interest to the media whose function is to report on the state of the world, or to editor-ialize about it, if only because growth to them means great-er circulation, a larger audience, and thus more advertising revenue. Virtually all institutions thus perceive their best interests to reside in promoting growth, in the economy and in the population as a whole.

THE DELETERIOUS EFFECTS OF EXCESS GROWTH

But a point has now been reached at which growth of population and of the economy are more and more clearly excessive. Shortages of raw materials, the increased costs of labor, taxes and social security benefits, and a host of other costs—material losses due to pollution, plus the expense of pollution control, are wiping out the gains derived from economic growth. Market saturation and inflation are fur-ther warning signals. So is the exodus from large cities to the suburbs that occurred between 1950 and 1970 censuses. What would life be like if the whole world were as densely populated as the center of New York City, and there were no suburbs for the middle class to flee to?

AN END TO GROWTH?

Despite continued advocacy of growth, the evidence is accumulating that growth may in fact be coming to an end. Since around the end of 1965, the Dow Jones average of thirty industrial stocks has essentially been hovering about a ceiling value. A variety of measures of economic activity show either slowing growth or a decline from previous peaks. As we have noted, the automobile industry, a bell-wether of the entire U.S. economy, shows a decreasing rate of growth. A rapidly dropping birth rate raises questions about expectations of growth.

If growth is slowing or has stopped, and institutions go on insisting that it can be expected to continue at a high rate, a variety of hazards is the result. Advertising may encourage investment in corporations or real estate, simply because a particular investment has been profitable in the past—and the investor who is misguided enough to believe this at a time when the value of the investment is declining may lose money. The hazards are still more serious if government also becomes a party to creating a false impression of growth—for example, by increasing the supply of money at a rate faster than is taking place in the country's real assets, leading to a drop in the value of money. For those on fixed incomes, this becomes a tragedy. But young people are also affected by inflation, since their incomes rise faster in "current dollars" than in constant dollars, sending every wage earner into too high a tax bracket as related to the actual buying power of those earnings. Speeding up the rate of increase in the money supply thus becomes a subtle way of raising income taxes without a rise in the tax rate, thereby depriving taxpayers of one obvious reason for disenchantment with the party in power. In effect, inflation becomes a means through which governments can collect a higher proportion of the earnings of their citizens.

Allowing prices on natural resources such as gasoline to stay low is another way of making an economy look "younger" than it really is. As we have seen, if the mechanism that would normally drive up prices as a resource became scarce were permitted to function normally, this would ensure that less would be used, and that serious efforts would be made to find a substitute resource and develop a new technology in its place. Government can put a floor under the price of any resource likely to become scarce, or it can pursue a laissez-faire approach, under which prices seek their own level. The trouble with this approach is the risk of price "wars" between marketing corporations which could keep prices low enough to discourage any motive on the part of the consumer to conserve the resource. A more far-sighted government policy in this situation would be to counteract the effect of the price war, thus ensuring a longer life for the supply of the resource. On the other hand, artificially keeping down prices on such resources amounts to a cosmetic device for masking a danger signal.

THE CHANGING CLIMATE OF OPINION

There is some evidence that individuals and institutions are coming to recognize the hazards of uncontrolled growth. In northern California, for example, the executive committee of Bay Area Governments took the step in 1965 of approving a plan to hold down the growth of population to 5.5 million by 1980—only a million more than the 1970 total for the region, but still representing an annual growth of 1.35 per cent over the fifteen-year period.

II

OBSESSION WITH PRODUCTION INSTEAD OF SERVICES

Many of the problems discussed in earlier chapters are caused by a basic structural defect in the United States economy—its over-emphasis on consumption of matter and energy, and its lack of concern with the provision of services. Although in production per capita the United States is the richest country on earth, a surprisingly small proportion of its wealth goes into services, as compared with several other countries. What does distinguish this country from all others, in addition to its high gross national product, is an extraordinarily high consumption per capita of matter and energy. A comparison of the U.S. with several other countries, arranged in order of energy consumption per capita (Table 1)[1] shows it to be 77 per cent higher than Sweden, which ranks third. In public expenditures for education as a percentage of national income, however, the U.S. ranks lower than Sweden, the Netherlands and Finland. Japan, with a per capita consumption of energy only 29 per cent that of the U.S. published 78 per cent of the number of book titles per million people; each of the eight other nations on the list published a far higher number of book titles, calculated on the same basis. Eight of the nine

TABLE 1   Consumption of energy and expenditures on services

Country	Rank in per capita consumption of energy	Energy consumption per capita in pounds of coal equivalents	Public expenditures for education as percentage of national income	Number of book titles published per year per million people	Persons per hospital bed
United States	First	24,568	6.3	388	124
Sweden	Third	13,913	8.1	958	68
Czechoslovakia	Fourth	13,911	4.5	625	98
Denmark	Sixth	13,060	6.0	1027	114
United Kingdom	Ninth	11,821	5.6	600	121
West Germany	Tenth	11,270	3.6	763	88
Netherlands	Eleventh	11,184	6.5	857	197
Finland	Sixteenth	9,149	6.5	1192	72
Switzerland	Twentieth	7,392	4.1	1324	89
Japan	Twenty-first	7,077	4.0	302	79

other countries had a higher proportion of hospital beds than the U.S.

Clearly, the allocation of resources in the U.S. is different from that in other developed countries. That far less money, in relation to a given flow of energy per capita, goes into services here, implies a different philosophy of national development. In this country, wealth is acquired through the utilization of matter and energy rather than through human labor. Because matter and energy are underpriced, the substitution can be carried too far. As a consequence, our priorities concerning some things that require a high expenditure of human time per unit of matter of energy input—education, culture, and care of the aged, for example—have been turned upside down.

ORIGINS OF OUR PRESENT LIFE STYLE

To understand the obsession with production and with the consumption of matter and energy in this country, it is necessary to uncover the historical roots of the phenomenon. During the nineteenth century, at a time when immense strides were being made in technology, the United States was better able than any other nation to exploit these developments because of an awesome abundance of natural resources, combined with a population large enough that economies of scale produced a very favorable situation for manufacturing and marketing, both domestic and international. The role played by natural resources in the development of the United States is not widely recognized, in part because several of those resources have ceased to be abundant. Whales, for example, were a mainstay of the New England economy, providing a great variety of materials ranging from corset stays and buggy whips to lubricants and illuminating fuel. The buffalo was similarly exploited;

TABLE 2 U.S. consumption of energy, 1850–70

Year	Resident U.S. population (millions)	National energy consumption in trillions of British thermal units (B.T.U.)				Per capita energy consumption	
		Mineral fuels. coal, oil, gas	Electricity from waterpower	Fuel wood	Total	Millions of B.T.U.	Pounds of coal equivalents
1850	23	219		2,138	2,357	102	7,758
1860	31	521		2,641	3,162	102	7,786
1870	40	1,095		2,893	3,988	100	7,611
1880	50	2,150		2,851	5,001	100	7,634
1890	63	4,475	22	2,315	7,012	111	8,473
1900	76	7,322	250	2,015	9,587	126	9,618
1910	92	14,261	539	1,765	16,565	180	13,740
1920	106	19,007	775	1,601	21,383	201	15,343
1930	123	21,506	785	1,455	23,746	193	14,732
1940	132	23,042	917	1,358	25,317	191	14,580
1950	152	32,552	1,601	1,164	35,317	233	17,786
1960	180	43,185	1,775	970	45,930	255	19,466
1970	203	64,565	2,879	794	68,238	336	25,633

numbering 80 million as recently as 1800, in some parts of the developing nation it provided food for settlers three times a day, every day of the year. More than anything else, however, the great advantage of the United States over all other nations was in its supply of energy. Other countries use less energy per capita because they have always been less well supplied. Their relatively greater investment in education and books has come about because they are older and more stable, rather than vigorously growing, and because they have needed a high level of technical competence in order to compete despite a shortage of raw materials.

In America, because it has always been possible to produce and consume great amounts of matter and energy, these wasteful tendencies have come to be perceived as synonymous with the "good life." How this attitude developed is suggested by the figures in Table 2.[2] As recently as 1880, the principal source of energy in the United States was fuel wood, and it was used in such quantities that the consumption of energy per capita in 1850 was already greater than it would be in Switzerland or Japan by 1970 (Table 1). Put another way, the fact is that after a century of diligent development in manufacturing and technological capability, Japan had not quite equaled the consumption of energy per capita reached in the United States by 1850, largely through cutting down trees.

Before the supply of wood could run out, however, coal was found to be abundant, and by 1865 it had come into heavy use. Long before the supply of coal could be depleted, once again, crude oil and gas were being exploited on a large scale. Thus, all through the history of the United States, a variety of fuel sources have been available in superabundance.

Contrary to a popular misconception, the most rapid

increase in the use of energy in the United States took place some time ago, rising from 111 to 180 million B.T.U. between 1890 and 1910 for an increase per annum of 2.44 per cent—as compared with 1.85 per cent over the twenty years between 1950 and 1970. Limitations on the supply of fuel wood now mean that it could not possibly replace the mineral fuels if they were to run out. The same is true of waterpower as a source of energy.

From a comparison of Tables 1 and 2, it is evident that the U.S. could cut its per capita consumption of energy in half and still maintain a living standard equal to that of Sweden—a country that in 1970 ranked third in its consumption of coal equivalents per capita! Perhaps this fact more than any other suggests the gap between the United States and the rest of the world.

The problems of resource depletion, pollution, market saturation, manpower glut, and inflation, are all merely symptoms of one basic defect: too much emphasis on consumption and not enough on services. Now, because of its great economic power, the United States is in a position to export this very defect to other countries, along with the misallocation of capital that is its concomitant. The result will very shortly be an overwhelming demand for resources.

In late 1972 the world's population was rising by 2 per cent annually and the per capita use of non-renewable resources was rising at a faster rate—the use of crude oil, for example, had been increasing by about 8.3 per cent annually. Both are serious problems; the growth of population, however, is the more serious over the long term. On the other hand, serious trouble is likely to develop before 1980 as a result of our misallocation of capital and resources. Supersaturated markets and spreading unemployment are likely to become serious in aerospace and air travel, in electronics, automobiles, and a number of other industries in all technically advanced countries. In the U.S. it is unlikely that either the aerospace or the automobile industry will see good markets beyond 1974. The markets for luxury housing and educators had been saturated by 1973. As has already been argued, the only solution to this problem is to divert a sizable part of the population now in manufacturing to service occupations.

The misallocation of capital is clearly due to a defect in our social philosophy—the positive value placed on a high rate of production rather than on efficiency in the use of resources, and the emphasis on raw power rather than sophistication in engineering systems.

Unless we break out of the rigid thought patterns that lead us to go on misallocating capital and depleting resources, we shall find ourselves locked into those patterns by the sheer magnitude of our investments. Once most of the cities in the world have made a commitment to freeways, for example, thereby affecting the design of the cities so that their density is too low to support mass transit systems, what will happen if mass transit is rendered necessary by a shortage of petroleum? The greater the number of nations locked into the same trap, the more difficult it will be to break out. If many nations coincidentally misallocate large blocks of capital, an international emergency—such as a world-wide shortage of petroleum—could mean a world-wide shortage of capital. Debt financing on a hitherto unprecedented scale would be a possible solution; but, as has already been noted, merely having the necessary capital available does not ensure that it will in fact be used for a particular venture. For example, the public has been known to vote against—and a U.S. President has been known to veto—expanding medical training facilities in the face of an acute shortage of physicians, as well as against strengthening school buildings where an earthquake hazard was known to exist, and against measures for reducing smog, even when deaths from emphysema and lung cancer were rising.

It is clearly dangerous to assume that at some moment in the future, public thinking will magically break out of its present pattern and provide a mandate for a realistic and rational allocation of capital. The U.S. electorate has repeatedly demonstrated an ability to vote against its best interest. This is particularly true when a present benefit (cash on hand) was being weighed against the cost of ensuring a future benefit.

Thus, the problem of capital allocation is political, as well as economic. Surely it is foolhardy to temporize in dealing with this aspect of the problem. The only prudent course is to begin educating the public in rational decision-making on such matters as resource depletion, efficiency in the use of matter and energy, unemployment and pollution. Inflation is in part a result of the over-emphasis on the consumption of matter and energy that characterizes our culture. If the world's population and the demand for resources both continue to grow, inflation can be expected to become extremely severe within a very few years. Cutting back on both population growth and the demand for resources per capita would lessen its severity.

Both renewable resources such as beef, ocean fish and shellfish, and non-renewable resources such as fossil fuels and metals, are likely to be depleted through over-exploitation by the early 1980s. Far too little thought has been given to the snowballing effect on critical resources as several of them approach depletion or exhaustion simultaneously. As one resource is nearly depleted, there will be sharply increased pressure on those that remain and that could be used as substitutes. The inflationary effects of this process are not pleasant to contemplate, nor are the crises that will ensue when the need for substitute technologies arrives too fast for our present slow-moving institutional procedures to cope with.

Probably the first such crisis will be in the area of energy. Up to the early 1970s, given a choice between coal and crude oil for home heating, people chose crude oil. Given the further choice between crude oil and gas, they chose gas. It happens, however, that neither of these preferred energy sources is very abundant on the earth's surface. The response thus far has been not to slow down the rate at which either is used but rather an effort to discover more of the same, or to develop a substitute. Few people have given much thought to the form such substitutes might take, or to what they would cost. The common assumption has been that substitutes will be available, in a convenient form and at about the same cost as for mineral fuel.

A rather desperate sequence of events could unfold, however, once gas, the most popular heating fuel, approaches depletion. A sharp conversion to oil will hasten its depletion—to be quickly followed by depletion of crude oil stocks, with a sharply increased pressure on stocks of coal and uranium oxide. The nation's coal supply will last for many decades; that of uranium oxide, for not more than a few.

A comparable scenario applies to food stocks—beginning with beef, followed by fish and shellfish and then by many kinds of fruits—and to construction materials, beginning with wood and certain metals.

The cure for depletion is not a frantic casting about for substitutes, but rather a variety of economic, political and demographic measures for cutting back on the rate of resource use: first of all, a curb on the growth and ultimately the size of the population, and an increasing efficiency in the use of resources.

The problem of pollution can best be dealt with not through emission control, but by using less matter and energy in the first place. Among the consequences of increased world-wide pollution will be climatic changes affecting the growth of crops. Obsession with production has led to international competition for markets, which in its turn hastens the depletion of resources. Because in certain product lines more goods are being produced than are needed, a significant number of workers in certain industries throughout the world are vulnerable to unemployment—either because the international market is saturated (as with airframes), or because of short-term changes in the relative position among competitors in several countries, (as with cars).

If more of the workers in all countries were involved in providing services rather than manufactured goods, unemployment would be less likely to develop as a result of international competition.

Many of the problems described in this book are caused by an obsession with production and consumption instead of services. If we had lower use rates for matter and energy per person, these problems would be far less severe. How much longer can we refuse to deal with them and still live in an approximately normal world?

The answer is: not long at all. Economic stagnation in the United States has been temporarily corrected by devaluing the dollar. But all that has been accomplished is to ship our problems abroad. The predictable result is some degree of economic stagnation for our trading partners—which will mean that their purchases from us must be curtailed. What happens then? With the U.S. continuing to sell less than it buys, the balance of trade can only worsen.

By 1972, the annual debt incurred by the U.S. government was already massive and increasing from year to year. How much longer the electorate would allow the government to lend financial backing to gigantic but unnecessary projects, simply to stave off unemployment, was unclear.

The sensible course would be for the government to shift its support from these to other gigantic but *necessary* projects for the same purpose: urban renewal, using architecturally novel structures such as Habitat, and mass transportation.

Finally, we can't go on as we are now because of the increasingly serious effect of pollution on both climate and public health.

WHAT CAN BE DONE

Believers in the necessity of economic "growth" have no choice but to support a new kind of "growth," with emphasis on efficiency in the use of matter and energy. There is no way to maintain high rates of economic growth, and thereby improve the lot of the less affluent half of the population, unless we maintain employment rates by deliberate shifts in the allocation of the labor force across sectors of the economy. We must either institute novel means of income distribution independent of jobs and accept more leisure, or sharply decrease our total production, thereby contributing to unemployment and producing economic stagnation. Once economic stagnation sets in, the pattern is difficult to break. The enduring effect of the Great Depression, as illustrated by the figures on per capita consumption of energy from 1920 through 1940 (Table 2), is evidence of this. It was not until 1950 that the figure rose significantly over that for 1920—and the long climb might have been even slower if the World War II had not hastened the process of economic recovery. Economic stagnation also tends to set in motion a snowballing process that can quickly suck in almost everyone in the population. It is noteworthy that, although a large part of the population were affluent in the early 1970s, very few people were at all affluent in 1933.

Specific steps can be taken to move the economy toward a more efficient use of matter and energy, and to orient the society as a whole toward service and quality. Tax rates should be operated on a sliding scale for all uses of matter and energy so that heavy users bear the main burden—for example, sales taxes on automobiles and on gasoline should be based on the horsepower of the engine. Since it is clear that private decision-makers cannot be relied on to deal wisely with national priorities, the government must restrict investment in some activities and promote it in others. Limits would be placed on the construction of service stations and the development of aircraft models, for example, while massive loans are made available for metropolitan transit systems, and for radical experiment with new housing materials and methods of construction. The government could also provide leadership in modernizing building codes, and in setting union rules for residential and office construction. Such steps would pave the way for urban renewal that is attractive, at a minimum cost per square foot. Large government loans ought also to be provided for experiment with new motors and sources of energy—notable with utility systems based on solar energy.

As a means of dealing with one of our gravest problems, Herman Daly has suggested the idea of resource depletion quotas, to be set for each basic resource during a given time period.[3] The legal right to consume the stipulated amount of a particular quota in each such period would be auctioned off by the government. As total depletion was approached, the units of quotas, small enough so that firms could bid for them, would go up in price—and thereby encourage conservation, since the prices paid at auction by the corporations would be passed on to their customers.

12

OVER-POPULATION

Most people have a notion of over-population as an ill-defined problem that will plague mankind at some time in the future. In fact, over-population has already occurred in many places, and can be assigned a cost which has many components, one of which is an increase in taxes. Logic, computer simulation experiments, and analysis of government tax data all bear out this contention.

The conventional wisdom can be stated in the form of a hypothesis, namely that *tax costs per capita decrease when there is higher population growth rate, greater population size, or greater population density.* It is important to notice that there are three components to this hypothesis. In a given urban area they might occur separately or together; for example, a very small but rapidly growing area will have a high population growth rate, but low population and low density. A city may have any of several combinations of population size and density. Houston has a large population (1.233 million) but a low density (2,841 per square mile). Baltimore has fewer people than Houston (906,000) but a higher density (11,568 per square mile). Cambridge, Massachusetts has a much smaller population than either (100,000) but a much higher density (16,187 per square mile). San

Jose, California, over the period from 1940 to 1970 had a population growth rate averaging 6.5 per cent annually, but still numbered only 446,000. New York grew by an average of only .19 per cent per year over the same interval, but has a population of almost eight million.

Growth rate, size and density may contribute in different ways to over-population, and in this chapter we shall distinguish among those ways.

RISING TAXES

In some cities, taxes have tripled in seven years. To put the various components of tax costs into perspective, Table 1 shows how taxes in various categories rose over a nineteen-year interval from 1950 to 1969.[1]

These figures bring out two important points. Even though the expenditures for national defense are enormous, the rise in this category has been less rapid than for almost anything else. More significant still is that most of the categories for which taxes have been growing very rapidly are related to population size, population density, and

TABLE 1   Per capita increase in taxes, 1950–69

| | Per capita tax costs | | Average annual increase per capita |
Category	1950	1969	
Higher Education	$7.27	$57.01	11.45%
Local Schools	38.79	166.60	7.97
Public Welfare	19.47	72.71	6.93
Hospitals	13.46	42.41	6.23
Health	4.34	16.47	7.27
National Defense	120.54	417.06	6.75
Old Age Insurance	4.76	160.29	20.33
Police	5.67	20.94	7.11

the age structure of the population. For example, the dramatic increase in expenditures for higher education can reasonably be attributed to an extraordinary increase in the numbers of people of college age, as compared to those in other age classes. Similarly, the tremendous increase in old-age insurance would appear to mean that there are many more elderly people than formerly, as compared with those of other ages. Police costs may have gone up because the population has become more urbanized, and the incidence of crime is greater in urban areas. That all these guesses are correct suggests a possible relation of tax costs to the growth rate and other characteristics of populations.

It is our contention, first, that the rate of population growth has an effect on the age distribution of the population, and second, that the age distribution has an effect on the tax cost per taxpayer. It can also be argued that age distribution has an effect on the demand for capital as related to the supply. Consequently, population age distribution is an important determinant of the rate of economic growth. Excessive rates of population growth constitute an important drag on economic growth.

In a population at any given time, there will be a certain abundance of people under five years of age as compared to those between five and fourteen, between fourteen and nineteen, and so on. At that same time, the population might have a high birth rate (50 births per thousand women in the same age category). Depending on whether this birth rate is high or low, in the following year the number of one-year-olds will be high or low as compared to the number of those older than one year. If there is a sequence of years with high birth rates, very young people will tend to be quite numerous as compared to people over twenty-five; or, conversely, a sequence of years with low birth rates would lead to a shortage of those under twenty-five. To put the same thing somewhat differently, if young people are added to the population faster than the rate at which old people die, there will soon be an age distribution with large numbers of young people as compared to the numbers of older people. Proceeding to examine the effect of age distribution on tax costs, we note that up to about the age of twenty-four, people in a developed society largely absorb taxes and capital, in the form of expenditures for school construction, teachers' salaries and other educational costs, and for welfare and health. They contribute little to the generation of taxes or capital. However, people after the age of twenty-four have completed their education and begun to share the adult burdens of generating the tax revenue and capital that keep society functioning. Thus, the ratio of people twenty-four years of age and younger to those who are twenty-five and older is one of several very sensitive indices of economic development. Knowing this ratio, we know a lot about the ease with which our society can meet its tax costs, and whether it can generate capital fast enough to provide jobs for all the young people who intend to enter the labor force.

COMPUTER SIMULATION RESULTS

There are difficulties, however, in making a case for the relation between population growth rates and economic variables, because of the great number of such variables. The use of actual data to demonstrate the relative effect on the economy of each of these variables, including population growth rate, would require methods of statistical analysis so complicated as to be unintelligible to many readers. In order to focus on the effect of population growth rates on taxes, with all other factors held constant, we can employ the method of computer simulation—that is, we can set up a mathematical model of a population and then put a computer to work on a series of questions, one by one, concerning the impact on tax costs per taxpayer of different population growth rates.

From this kind of inquiry, it becomes clear that the tax burden per taxpayer is remarkably sensitive to changes in the rates of population growth.[2] At a zero growth rate, for example, the ratio of those who are primarily tax consumers to those who are primarily tax producers is .33; at a growth rate of one per cent annually the ratio becomes .45; at a growth rate of 2 per cent, .59; and at a rate of 3 per cent, .74. This means that if a population whose size had been constant began growing by one per cent a year, the resultant change in age distribution would increase the tax burden per taxpayer by 100(.45/.33), or to about 136 per cent of what it would have been with no growth. For a population with a growth rate of 2 per cent, the figure would rise to 179 per cent, and for one growing at a rate of 3 per cent, to 224 per cent of that at zero growth.

Such figures might well be borne in mind by citizens' groups confronted with the arguments of a county board of supervisors, or a developer, that accelerating the growth of population in an urban area can broaden the tax base and thus reduce taxes. The counterargument could be put forth that a principal cause of population growth in a rapidly expanding urban area is immigration, which may actually improve the age distribution from a tax standpoint, since many of the immigrants will be young adults who have just completed their education. This is true; however, these young adults will shortly have children whose education is what turns out to be so expensive.

Computer simulation also suggests that a high population growth rate may put a demand on a government for taxes so great that the electorate simply rebels, refusing to support bond issues for education. From computer simulations such as those, we can go on to interpret recent changes in the age distribution of the U.S. population.

THE TAX IMPLICATIONS OF RECENT CHANGES IN THE U.S. POPULATION AGE STRUCTURE

Changes in the U.S. population since 1930, with growth per annum for each age group, are shown in Table 2.[3]

TABLE 2   Growth of U.S. population, 1930–70

| | Population (in millions) | | | | | |
Ages	1930	1940	1950	1960	1970	Growth per annum
Under 5	11.5	10.6	16.2	20.3	17.2	1.0%
5–14	24.7	22.5	24.4	35.5	40.7	1.3
15–24	22.5	24.0	22.2	24.0	36.2	1.2
25–34	19.0	21.4	23.9	22.8	25.1	.7
35–44	17.2	18.4	21.5	24.1	23.2	.8
45–54	13.1	15.6	17.4	20.5	23.2	1.4
55–64	8.4	10.6	13.3	15.6	18.6	2.0
65–74	4.7	6.4	8.4	11.0	12.4	2.5
Over 74	1.9	2.6	3.9	5.6	7.6	3.5

This rather strange profile clearly shows the effects of the Depression. Several years marked by a low birth rate show up as a smaller than normal increase in the population between the ages of 25 and 44. Even a cursory examination of the table indicates obvious implications of such changes for tax rates, supply in relation to demand for capital, and rates of economic growth. For example, in 1940 and 1950, following the Depression, the numbers in the expensive age bracket—from 5 to 24—were not large as compared to those in the vigorous tax-generating and capital-generating groups between the ages of 25 and 44. By 1960, however, and even more by 1970, the numbers between the ages of 5 and 24 had become very large in relation to those 25 and 44 years old. In 1940, the ratio was 1.17; in 1950, it had dropped to 1.03, rising to 1.27 in 1960 and to 1.59 in 1970. It is noteworthy that in 1930 this ratio was 1.30. Since the effects of these age distributions tend to show up in the economy after a lag of some years, the high ratio for 1970 will probably be reflected in the economy of a few years hence.

The tax and capital implications of changing age structure are brought out more clearly if we group ages according to their probable share of the tax burden, as in Table 3.

TABLE 3   Age structure and the tax burden, 1930–70

	Population (in millions)				
Age	1930	1940	1950	1960	1970
Under 25 (Tax Consumers)	58.7	57.1	62.8	79.8	94.1
Over 64 (Tax Consumers)	6.6	9.0	12.3	16.6	20.0
Total Consumers	65.3	66.1	75.1	96.4	114.1
25–64 (Tax Producers)	57.7	66.0	76.1	83.0	90.1
Consumers Per Producer	1.13	1.00	.99	1.16	1.27

This ratio of consumers per producer is somewhat analogous to overhead in a business, calculated per unit of productive potential. It operates as a drag or brake on growth in the economy, since the money used to support non-producers cannot be spent elsewhere, either to improve the quality of life or to invest in new ventures. This being true, we would expect to find a relation between this ratio at the beginning of a decade, and the rate of economic growth throughout the decade. Table 4[4] allows us to test this assertion.

TABLE 4   Percentage growth in GNP

Year	Consumers per producer	Percentage growth[a]
1920	1.18	−20
1925	1.17	−20
1930	1.13	38
1935	1.05	113
1940	1.00	34
1945	1.04	40
1950	.99	27
1955	1.05	36
1960	1.16	42
1965	1.27	?
1970	1.27	?

[a]During five year period beginning five years later.

The table has been arranged on the assumption of a long lag between the time a particular consumer/producer ratio goes into effect and the disclosure of its economic consequences. One explanation of this lag is that if there is a problem of unemployed capacity, a high birth rate may increase the aggregate demand and thus the measurable growth in the gross national product. This occurs only up to a full employment limit, however, and is consequently a short-run phenomenon. According to the record in this table, under consumer/producer ratios of 1.17 no consistent relation between the ratio and economic growth can be discerned. The reason for this is that population age structure is not the only important determinant of economic growth rates: wars and other phenomena are also clearly important. However, as the age structure interacts with a number of other factors, its role in determining economic growth rates is significant. It may thus be a portent that the consumer/producer ratios in 1965 and 1970 were extraordinarily high as measured against the historical record.

LONG SWINGS IN THE LABOR FORCE AND THE ECONOMY

The total annual number of births in the United States did not begin to drop until 1962. The ratio of the population under 25 to those of 25 and over reached a peak in 1966, then began to decline as birth rates plummeted. But this high ratio will continue to have a depressing effect on the economy for many years, in part because of the inability of young people to find work. Thus, while education costs will slacken somewhat, welfare costs will compensate.

Statistics on the rate of teen-age unemployment provide us with important insights into what the immediate future holds. The educational or welfare burden of these young people must be borne in some fashion, either by the state or by parents. In either event, such teen-agers have little motive for a high productive rate. Much attention has been given to the resultant long swings in birth rates, unemployment and the labor force,[5] called "Kuznets cycles" after Simon Kuznets, the economist chiefly credited with pointing them out.

Educational costs per capita are now so high that in many areas services are being cut back or even discontinued. This sets up a vicious spiral, since educationally deprived youngsters have a harder time finding jobs. Severe unemployment among young people is reflected in weakening retail sales, which further contribute to the downward spiral of an economy.

POPULATION SIZE AND PER CAPITA COST OF GOVERNMENT OPERATION

Up to this point, our discussion has considered only population growth rates and their effect on age structure and economy. We must now consider the separate effect of population size and density.

Many inefficiencies of operation follow from gargantuan size, whether in a city, a country or any other institution. This is demonstrable from the financial records of local government. Since details vary considerably from city to city, even cities of the same size, generalizations about them are best derived by pooling data on cities in six different categories of size. Tax expenditures per capita for local governments in each of the six categories are given in Table 5, along with comparable data for New York City.[6]

Some caution should be exercised in interpreting such a table, since it tells us nothing about the quality of the services provided or required. Certain broad qualitative patterns do emerge, however, on which many authorities are agreed. So far as tax costs alone are concerned, the optimum city size would appear to be somewhere under 50,000 inhabitants. When other economic data are included, the optimum may be higher; some experts believe it to be around 250,000—a size that certainly allows for such amenities as a concert hall and symphony orchestra.

Analysis of the figures here would clearly suggest, at any rate, an optimum at some intermediate size. For some services such as education and highways, the costs per capita are significantly greater at very low densities as compared with higher densities. Other costs, such as public welfare, police and fire protection, rise very steeply as the population goes above 250,000 (Chapter 6).

Thus, a developer who suggests that an urban area be expanded so as to decrease tax costs per capita will be correct only if the city has considerably fewer than a quarter of a million inhabitants. Moreover, even in the lower categories of population, any increase in size may raise rather than lower tax costs if that increase entails a low density, with

TABLE 5 Per capita tax costs in local government

Category by size	Less than 10,000	10,000– 24,999	25,000– 49,999	50,000– 99,999	100,000– 249,999	250,000 or more	New York City (7.9 million)
Direct General Expenditures	$269.00	$232.00	$232.00	$246.00	$273.00	$351.00	$891.00
Education	151.00	133.00	131.00	137.00	146.00	153.00	188.00
Public Welfare	12.00	10.00	10.00	11.00	14.00	28.00	192.00
Police	5.70	5.20	6.00	7.50	9.60	19.00	52.00
Fire	1.40	1.80	3.20	4.80	6.80	11.00	22.00
Highways	40.00	30.00	25.00	22.00	22.00	21.00	22.00

sprawl, so that services must be extended over great distances.

From a purely economic point of view, in other words, over-population may already be a fact for many cities. Since population densities throughout the world are expected to increase by an average of at least two and a half times their 1970 level before growth stops, the conclusion is a provocative one. Average densities are not in fact the central problem: as these increase, the density of cities will increase still more, since rural to urban migration occurs in all developing societies. We thus have a foretaste of coming diseconomies of scale for many cities by looking at those that already exist in our largest cities. And that foretaste is alarming.

The data for New York City raise disturbing questions about very high-density living. Is it a way of life we can afford? Is there any way we can avoid the tremendous costs associated with it? And most disturbing of all, can it be that no nation can support more than a limited proportion of its population under conditions of very high density, so that to exceed this proportion would be to destroy the economy?

One fact is inescapable. Not only is high-density living very expensive; it is also very unpleasant. The most densely populated urban areas in the United States have typically undergone a large net exodus of population over the last two decades. Over 9 per cent of the population of Manhattan, for example, left between 1960 and 1970. During the same decade, this pattern occurred in Baltimore, Boston, Cambridge, Camden, Chicago, Detroit, Jersey City, Brooklyn, Newark, Philadelphia, St. Louis, San Francisco, Trenton and Washington. Planners who assume that there will be much greater population densities in the future ought to determine whether anyone will put up with them voluntarily.

Our conclusion must be that low-density sprawl and high-density concentration are both too expensive: the optimal situation is somewhere in between. Compact cities of about 250,000 people, with a density somewhere between 4,000 and 8,000 per square mile, represent the most efficient life style. To suggest how such an optimum compares with the actuality, here are the population densities per square mile for some U.S. cities in 1970:[7] Manhattan, 67,808; San Francisco, 15,764; Washington, D.C., 12,321; New Haven, Connecticut, 7,484; Richmond, Virginia, 4,140; Houston, 2,841; Las Vegas, 2,438; Oklahoma City, 576.

Nations that double or triple in population—as it now seems many nations will—have unfortunately not been inclined to accommodate much of their growth in new towns. Thus, we must expect our largest and densest cities to grow still larger. If we suppose the age structure of an entire nation to include a disproportion of very young people, as a consequence of excessive population growth rates, and an excessive population concentration in large cities with diseconomies of scale, the combination may conceivably be such as almost to prevent economic growth—which may in fact be the present situation. And if the bulk of an entire population lived in cities as densely populated as New York, the total social cost per capita, averaged over the entire nation, might be such as to cause a major economic collapse.

This line of speculation suggests that there may be an optimum range of size not only for cities, but for whole nations as well—and that to exceed that optimum may lead to economic ruin. We must either let the market place regulate numbers of births, as at present, or offer disincentives to reproduction by means of tax legislation, welfare reform, and distribution of birth control devices and information.

The difficulty with leaving the regulatory function to the market place, given the long time involved, is that a great deal of hardship may be experienced by large numbers of people while they wait passively for the market place to deal in its cumbersome and insensitive fashion with a problem that could have been handled far more rationally by government.

One possibility that has not been explored is truthful propaganda—for example, a pamphlet distributed by the government in high schools, colleges, supermarkets, and doctors' offices, entitled *What a Baby Will Cost To You,* and dealing realistically with the following:

The immediate costs of having a child;

The future costs of educating a child, and the economic prospects for a child in a crowded world if it is not highly educated;

A projection of the effect of inflation on all costs of raising a child, assuming increase in price of stock resources as a result of increasing depletion and the inaccessibility of remaining resources, and increasing prices for renewable resources as a result of scarcity in relation to demand;

The economic prospects for a divorced woman with children, and a projection of probability of divorce;

The effect, for young parents, of having a child on the probability of getting an adequate education;

The prospects, for young parents, of finding work in a labor market diminished by distortions in the age structure of the population.

It may well be that if the government or some private concern began distributing realistic literature of this sort in great quantities, the problems of birth rate and over-population would be dealt with, since only very economically secure young people would wish to have children.

THE CONVENTIONAL WISDOM

In the modern world, information has become an important determinant of the way social, economic and political systems function. To a considerable extent, the public conception of reality is managed and determined by the government and other large institutions. Few individuals, and few organizations, are equipped or motivated to assemble the data needed for an understanding of the world. If the government or other institutions see fit to conceal, distort or delay the release of information, there is very little that most people can do about it. Indeed, very few people have a clear notion of how to go about finding out the truth behind what is presented to the public as the conventional wisdom.

An important impediment to perceiving the true state of the world is being set up by the official dissemination of information. A more subtle problem is the use by the government of indicators which become part of a feedback control system leading to economic instability. The following are typical of news items designed to indicate that the economy is healthy:

Issue 1: Rising Gross National Product. The Annual Report of the President's Council of Economic Advisers forecast that during 1972 the gross national product would rise 9.5 per cent, from $1.047 trillion to $1.145 trillion. "Real" or deflated GNP would rise about 6 per cent, with only 3¼ per cent of the total rise being due to inflation. This was said to represent a great improvement over 1971, in which real GNP rose only 2.7 per cent, and inflation increased 4.6 per cent.

Issue 2: Unemployment. Business literature tends to play down the significance of an overall unemployment rate of about 6 per cent for the reason that the unemployment rate among married men is only a little more than half that: about 3.3 per cent. The argument is often presented that the national rate of unemployment is misleading since the problem is concentrated among groups of marginal significance to the economy. Thus, among teen-agers the unemployment rate is about 20 per cent of those in the labor force, while the overall rate is 6 per cent.

Issue 3: Economic Indicators. That the government's composite index of leading economic indicators has been rising almost continually since October, 1970, is taken to mean that the economy will be functioning at a higher rate in future months.

Issue 4: Retail and Automobile Sales. Month after month over a long period, these sales figures have been significantly higher than for the corresponding month the previous year. That in typical recent months overall retail sales have been about 11 per cent higher than in the corresponding month the previous year, would seem to indicate beyond any doubt that the economy is healthy and expanding.

Issue 5: Taxes. Taxes are to be decreased to stimulate spending.

Issue 6: Government Spending. Government guarantee of loans on the Lockheed airbus, and increased government spending on space and defense, are expected to stimulate employment and the economy generally, thus guaranteeing economic health from now on.

Concerning each of these issues, other information would shed an entirely different light. For example:

Issue 1: Rising Gross National Product. What matters to us as individuals is not the deflated gross national product, but the deflated gross national product *per capita*. When such a correction is made, the gains look considerably less impressive. Thus, there was only a 2.7 per cent increase in real GNP in 1971; after making a correction to allow for population growth, the individual standard of living turns out to have increased by only about 1.7 per cent.

But this is only the beginning of the problem. GNP measures the total national output of goods and services. Some of the items being measured do not so much contribute to the quality of life as indicate that it is deteriorating— such as the costs associated with the control of pollution, crime and venereal disease, or with highly processed food such as TV dinners. It can be argued that these components of the total national product would be less high if certain aspects of life were more salubrious.

Furthermore, a constantly rising GNP may mask serious difficulties in large sectors of the economy. Much of the growth in recent years has been due to bureaucratic proliferation rather to an increase in real services or an increased tempo of production. By the early 1970s many measures of economic activity were at a lower level than they had previously been (Chapter 5). The number of housing construction starts, a particularly revealing index of future economic activity, was down. To say all this is not to argue for a high level of manufacturing, but rather to point out that GNP measures may not tell the whole story. By the end of 1971, in fact, manufacturers in the United States were operating at three-quarters of capacity—a deficit that was apparently not being made up in delivered services, which were being cut back noticeably in many places. A sharp increase in the GNP certainly doesn't, for example, indicate improved medical services, but rather a significant increase in their cost.

Issue 2: Unemployment. Even though very high unemployment statistics may be confined to certain segments of the population, such as teen-agers, such groups are not unimportant to the rest of the economy: Teen-agers, after all, will be adults some day. There is evidence that when economic pressures on the young are more severe than on middle-aged people, birth rates drop. In 1968, births to women between 25 and 29 years old were 71 per cent of what they had been in 1960, and births to women between 20 and 24 years old were only 65 per cent of what they had been in 1960. The greater drop in births to the younger group may be a reflection of the greater economic hardship they had experienced at a time when affluence was giving way to greater economic stringency. Business literature consistently fails to link the phenomena of efficiency and unemployment. Business magazines regularly predict how the U.S. will win out over foreign competition by increasing efficiency. But this victory always means more work out of the same or fewer laborers, so that to win the battle for efficiency is to lose again in the battle against unemployment.

Issue 3: Economic Indicators. The composite index of leading economic indicators is heavily influenced by stock price averages, which in turn are heavily influenced by the composite index. The result is a circular feedback system, which leads to instability in the economy. The index is also heavily influenced by building permits, which may in fact be an indication of overbuilding; when this occurs in a particular area, the sequel will be a sharp fall-off in the rate of building, and perhaps of all economic activity. It is possible that in Hawaii (Chapter 9) the overbuilding itself, by making the islands less attractive for tourism, contributed to the decline in business activity.

To be reliable, the composite index should include more leading indicators of firm intentions in such industries as coal and machine tools, which in turn produce the input for cars, buildings and appliances.

Issue 4: Retail and Automobile Sales. These figures are expressed in terms of current dollars rather than physical quantities. They must thus be examined carefully, since inflation can make it appear that much greater quantities are being sold from year to year even when there is little growth in the actual volume of goods being sold. For example, large dollar increases registered from year to year in the business of eating and drinking establishments may simply reflect a rise in food prices, as Table 1, giving sales figures for eating and drinking places, corrected for inflation, suggests.[1]

TABLE 1   Sales in eating and drinking establishments

Year	Sales in current dollars	Value of consumer dollar (1967 = $1.00)	Sales in 1967 dollars
1965	$20.2 billion	$1.058	$21.4 billion
1969	27.0 billion	.911	24.6 billion
1971	31.1 billion	.824	25.6 billion

Issue 5: Taxes. Although decreasing taxes to stimulate spending sounds at first like a great idea, it has far-reaching implications for a federal government already faced in 1972 with a deficit of about $36 billion. It is noteworthy that after President Nixon announced his budgets for 1971 and 1972—both of which had large deficits—the free market price of gold rose sharply. One possible interpretation of

this development is that foreign speculators foresaw an even larger deficit than had been planned for—and with it the prospect of a return to a high rate of inflation and a decrease in the real value of the U.S. dollar as against other currencies. Consequently, some speculators would be prepared to gamble that as a way out of a deteriorating international trade situation, the United States might choose to devalue the dollar—reducing its value, say, from 1/38 to 1/45 of an ounce of gold, through the expedient of raising the official U.S. price of gold from $38 to $45. Under some such expectation, beginning in late January 1973, gold speculators rapidly bid up the price of gold from $43 to $123 a troy ounce.

Issue 6: Government Spending. As has already been argued, government spending on space, defense and guaranteeing such projects as the Lockheed airbus is bad economic news. Without these expenditures, unemployment might be higher by at least 300,000 to 400,000 people. Thus such moves only postpone the inevitable and necessary shift of the economy away from manufacturing toward the provision of services. What we have is a whole set of "make work" projects designed to keep the level of manufacturing unnecessarily high.

The "good" and "bad" versions of these six issues reveal a conventional wisdom that fails to look ahead, to deal with such phenomena as market saturation and unemployment as an outgrowth of efficiency, or to explain today's news in terms of fundamental economic processes. Behind this failure looms the unacknowledged truth that there is no way to keep the economy healthy and growing except by limiting growth almost entirely to the service sector.

In thinking ahead, it is important to consider what the ultimate consequences of a particular process will be if it is continued indefinitely. This pattern of thinking leads to a concern with asymptotes—limiting processes as applied to the number of cars or television sets or jet aircraft that can be sold each year, or to the degree of pollution human beings can withstand.

An index of the way people in other countries view our economic status may be found in the purchases of gold and silver on the free markets—which are recorded daily in the *Wall Street Journal*—and also in the stock prices of such companies as Homestake Mining, Dome Mines and American South African, which are listed on the New York Stock Exchange and thus carried in most daily newspapers. Sudden sharp increases in the value of such stocks may indicate that speculators in Europe or the Middle East believe the real value of the U.S. dollar is dropping too rapidly because of inflation, leading them to gamble that the dollar price of gold will rise, and accordingly to buy up gold in order to benefit. The price of gold has historically been a reliable index of coming political and economic instability.

Unemployment is also a useful index of the state of an economy, even though most government officials deny that this is so. In fact, the changes from month to month in the proportion of the labor force who are unemployed tell us a great deal. If unemployment keeps rising, we can be sure that once it has passed a certain threshold, sales will begin to drop. And if unemployment figures remain high over a period of months despite persistent efforts at a cure, there is reason to believe that the number of unemployed is not being realistically measured—specifically, that people leaving the category of the unemployed are being replaced in that same category by others who have been without jobs but had not been so classified (Chapter 7). This seems to be what happened during the second half of 1971.

In observing the economic scene, one should train oneself to look for the connections between developments: between a high percentage of unemployment among teen-agers and a dropping birth rate, for example, or between the President's budget announcement and increasing gold prices. Increasing sales of foreign cars in the U.S., combined with decreasing sales of U.S. manufactured jet aircraft to other countries, are factors leading to a deficit in the U.S. balance of trade. Increasing that deficit means in turn that more U.S. money will be sent abroad and less foreign money will come in. The consequent accumulation of U.S. dollars in foreign banking systems means that ultimately these banks cannot buy up U.S. dollars fast enough to avoid having them glut foreign money markets, bringing a drop in their value—which in turn leads to the run on gold by speculators.

Business newspapers and magazines, by helping to generate optimism about the economy, thereby improve their prospects of maintaining subscriptions and advertising revenue. What motive has an advertiser to increase his expenditures if the economy appears to be declining? Because all reporters, both in the government and in the news media, have such motives for presenting the facts as favorably as possible to their own immediate prospects, the sophisticated reader will allow for this tendency in interpreting the news.

An interesting example of the impact of government statistics on the system they are supposed to be measuring is the composite index of leading economic indicators. That it is taken seriously by investors has been repeatedly demonstrated.

On Monday, September 25, 1972, for example, prices on the New York Stock Exchange continued a generally downward trend that had begun fourteen trading days earlier. During that period, the Dow Jones industrial average of thirty stocks had declined from about 970 to 936. In response to an announcement on the morning of September 26 that the composite index of leading economic indicators had risen very sharply in August—by 2.2. per cent over July—the Dow Jones average promptly rose about ten points.[2] Even more important to an understanding of how the composite index functions, however, is the phenomenon of feedback, which calls for a more extended explanation.

CAUSAL FEEDBACK SYSTEMS

Around the beginning of World War II, engineers and scientists became interested in the positive and negative aspects of circular causal feedback. A military example would be the radar sensing devices used to aim anti-aircraft guns. If such a gun had been aimed too far to the left to hit its target, the radar and the negative feedback control system attached to it would cause the gun to move to the right. If the gun had been aimed too high, the negative control system would lower it. In general, such negative feedback controls increase accuracy by correcting an error through a *change of sign* from positive to negative or from negative to positive; that is, a thing that is going too far one way is made to go *the other way*. By this means, errors in performance are constantly being minimized or reduced. A more homely example occurs in drawing a bath: if the water gets too hot, one turns down the hot water faucet and turns up the cold. Here, once again, the central element in the control system is a *change of sign*—in this instance, from hot to cold. Such feedback loops can be found everywhere: in engineering, physics, chemistry and biology, and also in social systems. In the stock market, for example, if the average price of shares runs too high in relation to the earnings of the corporations represented by the shares, prices tend to drop. As such a drop becomes probable, the ratio of stock prices to the earnings per share of the corporations goes up.

Such causal feedback systems are described as "circular" because situations of this kind operate to produce a never-ending system of circular cause and effect.

But such systems need not always be *negative,* even though they are so in nature. *Positive* feedback control systems tend to blow up and to become unstable. Since nature does not deal kindly with blowing up or with instability, such systems tend to be weeded out in the process of natural selection. The world of men, however, sometimes inadvertently produces a positive feedback system—one in which the bigger a thing is, the greater its tendency to become still bigger. Such things tend to become at once gargantuan and unstable.

An example is the highway system. Since taxes associated with the operation of automobiles go into the Highway Trust Fund, and must be used for construction of more highways or freeways, the number of miles of freeway must continually increase. This is a positive feedback system, with no built-in check as in negative feedback control systems. Until some major change eliminates the positive feedback, such systems go on gyrating more and more wildly out of control. Thus, the freeways will continue to grow until people come to realize that rather than relieving congestion, freeways only foster it, hastening the day when all of America is paved over—unless new legislation diverts moneys from the Highway Trust Fund to mass rail transit.

This same phenomenon of positive feedback loops is involved in the composite index of leading economic indicators. During 1971 and 1972, that index was calculated from just eight indices of economic activity. An index computed from so small a number of items is unusually sensitive to rather large percentage changes, month by month, in any one of the eight items. One of those items was an average price for 500 common stocks—and throughout much of the period, the month-to-month increase in the composite index was due in large part to an increase in that item.

Examples of the exact language in which the increases in the composite index were announced—in each instance cited here by Mr. Passer, Assistant Secretary of Commerce for Economic Affairs, as quoted in the *Wall Street Journal,* are of interest:

> Mr. Passer also said that since last August, a month before the General Motors Corp. strike began, the leading indicators have risen 2.3 per cent, "with new housing permits and common stock prices showing the most pronounced vigor" (February 25, 1971). The February index was up .6 of one per cent from January. Passer said that "most of the changes were moderate, except for the decline of 0.4 hours in the manufacturing work week and the increase of 4 per cent in common stock prices" (March 29, 1971). "The strongest increases were recorded in the manufacturing work week and common stock prices," Passer noted (April 28, 1971).

The point of these examples is clear: during a period when the economy was very slowly dragging itself out of a recession, the dominant influence on the composite index of leading economic indicators was the average of common stock prices. Still more to the point, however, is that the composite index of leading indicators is watched carefully by investors in the stock market where a rise in the composite index is taken as good news, and becomes the ground for buying stock. Thus what we have here is a *positive* feedback control system: that is, the composite index goes up because the stock averages go up, and then the stock averages go up because the composite index goes up! Great optimism can be generated, in short, for no better reason than that there has been great optimism in the market in the past: the system simply feeds on itself.

This positive feedback loop operates not only when stock prices are rising, but also when they are on the decline. When a market that has been stimulated in part by a rising composite index finally stops rising, for lack of further demand, the drop is exceptionally fast and deep, again because it is soon going down hand in hand with the composite index. In a situation where the stock average has been dropping continuously, and hard, for three months, a further drop would be precipitated as investors took note of the composite index of leading indicators, which would also be dropping fast and hard—even though by then, a further stimulus in that direction is the last thing anyone could want.

As has already been pointed out, many forces at work in our society tend to veil or obscure the true state of the economy at any given time. This being so, it becomes all the more deplorable that such a mechanism should have been built into the government's system of reporting economic statistics. To remedy this situation, two important changes should be made as soon as possible: the composite index of leading indicators should be computed from a greater variety of economic statistics—twice as many would be desirable—and the index of stock prices should not be used in computing it. Even without the composite index, the market is likely to have a positive feedback so long as trends in the recent past determine expectations of the near future. Stabilizing speculation would produce a negative feedback. Thus, speculation can be stabilizing or destabilizing according to the way the recent past influences expectations of the future, regardless of the role of the index.

THE MEDIA AND THE SYSTEMATIC
DISPENSATION OF OPTIMISM

The government is not alone in contributing to a mood of false optimism when in fact the economy bristles with warning signs. Much of the blame must be shared by the media, particularly business newspapers and magazines. Their motives are presumably to avoid "rocking the boat," which would bring criticism or reprisal from advertisers, and also to build business confidence, which produces more support for the media in the form of subscriptions or advertising revenue. But this reasoning is short-sighted. It contributes to a climate of unreality within the government and business management circles which makes it very difficult to generate support for aggressive action in dealing with emergencies.

Throughout a study of predictions made in business, industrial and investment literature begun in 1957, I have noticed that such predictions are never accompanied by a table of statistics indicating the past accuracy by their particular source or sources. Readers' attitudes toward economic predictions might very quickly become far more sophisticated if those predictions were accompanied by such an analysis. The forecasts then would have a real value. Much economic, business and market "news" is typically defective in four categories:

1. A widespread tendency to report on some small component of the entire economic situation, without showing how that component is related to the system as a whole. Conversely, important components are omitted when the entire system is discussed. The media in general have failed to present a holistic view of the economy. For example, analysis or predictions of unemployment for 1973 and later years have almost invariably failed to mention anything

about the demographic structure of the population. The significance of the 17.5 per cent unemployment rate among teen-agers in December, 1971 went largely unnoticed. A useful estimate of future unemployment would involve both projections of the rate at which new workers will be trying to enter the labor force, and a specification of those parts of the economy that will be growing fast enough to provide jobs and thus to prevent a rise in unemployment. This failure to view the system as a whole is also reflected in editorials and news analysis. Some analysts, for example, have noted an alarming rate of unemployment in one part of the country, only to conclude that there was no reason for continued concern, because large numbers of unemployed workers were emigrating to other states. A thorough analysis of this situation would include an inquiry into whether this shift would simply add to the rate of unemployment in the states they migrated to.

Since the international economic system is full of reciprocal causal feedback loops, any report of difficulties in the European economy that does not explain its probable effects in the U.S. is incomplete.

2. The omission of quantitative justification for assertions made. We are told the probable value of the gross national product for the next year, or the probable rate of decrease in the value of money, or the level of unemployment with no indications of the basis for these assertions. With over a dozen fairly complex computer simulation models of the U.S. economy in operation, in the absence of any specific statement to the contrary, it may be reasonably assumed by an informed person that predictions are based on one of these. Certainly, there is little reason to accept predictions made *on any other basis.* But unless he is told the origin of the predictions, the reader is put in the position of playing Russian roulette. Business media clearly are obliged to specify the origin of "authoritative statements."

3. The reliance of many business and financial journals on "expert opinion," once again without an adequate account of the basis for that opinion. Unless we know whether it represents a guess or set of guesses based on no analysis at all, or five minutes of figuring on the back of an envelope with a slide rule, or months of computer analysis, we are scarcely in a position to decide what to make of the opinion.

4. An alarming tendency toward de-emphasizing negative news. A typical example is the optimism over the devaluation of the U.S. dollar, which was sure to help our sales abroad—though the media typically failed to point out that it would hurt foreign sales here, and that in the long run other countries would therefore have less money with which to buy our products.

During the 1958–59 recession McGraw-Hill Inc., publishers of *Business Week,* used as the theme of a campaign to sell advertising in their approximately thirty business magazines, "The Year Advertising Helped to Kill a Recession"— while at the same time, McGraw-Hill executives were ordering the company's own advertising budget slashed by one-third. Clearly, optimism was for external purposes only.

On November 5, 1972, the *Los Angeles Times* published an article headlined "Despite Slow Sales, Boeing Still Expects to Soar With the 747." Midway through the article Malcolm Stamper, president of Boeing, was quoted as forecasting that the 300 billion revenue passenger miles flown in 1970 by commercial airlines in the non-Communist world could triple by 1980. This remarkable prediction passed unchallenged, even though to triple a business in a decade is very rapid growth indeed for any enterprise. By failing to make such an obvious point as that a 5.4 per cent annual rate of growth in revenue passenger seat miles by domestic scheduled air carriers would appear to be more reasonable than Stamper's prediction of 13.7 per cent, the *Los Angeles*

Times helped foster an unquestioning readiness to accept projections of very high rates of economic growth.

Business magazines often base such projections on inadequate analysis of the available data. An amusing example was the front page of *Barron's* for January 8, 1973, headlined "1200 on the Dow," followed by: "That's a modest expectation for 1973, says our year-end panel." In fact, the Dow Jones Industrial Average peaked at about 1060 on January 11, and had dropped to 783 by December 5, 1973. There had been a long and continuous decline of the stock market throughout the first six months of 1973, with a tremendous erosion of the value of stocks. A statistical measure which is widely-based is the Standard and Poor index of 425 industrial corporation stocks. This index lost about 19 per cent of its value in the first 11 months of 1973. Since this represents the pension funds, life insurance funds, and investments of a multitude of average people, this is a very serious phenomenon. How can a society be mobilized to deal with its economic problems in the face of such persistent denials that we even have any?

Mobilizing the public behind a commitment to the solution of these problems will not be possible unless both the media and the government undertake an intensive campaign to alert and educate the public to the urgency of the situation. This is no time to take refuge behind pleas of "Let's not rock the boat, it might be bad for business." It will be very bad for business if the system breaks down to a point beyond repair, as angry mobs of unemployed young take over!

THE ROLE OF TIME LAGS IN VEILING FUTURE DISASTER

In an ideal world, all problems would be promptly recognized and dealt with. In the real world, however, years may elapse before a developing problem is recognized. Another ten years may elapse before remedial legislation is passed. Even after legislation is passed, years may go by before corrective action will have had a demonstrable effect. If by 1973, politicians were to have recognized a major economic problem in the rate at which young people were trying to gain admittance to the labor force, the difficulty would still be that at this or any other time, the next twenty age classes to seek admission to a future labor force would already have been born!

Although this is an extreme example, the effect of long time lags in economics has been increasingly recognized. Much of the evidence comes from the efforts of government to manipulate the economy. For example, if inflation is getting out of control, the government may cool the economy by a cut in spending; or on the other hand, if the economy isn't growing fast enough, the government may adopt the opposite policy, of spending money on a great variety of projects, to the point of operating at a deficit and adding to the national debt. Besides "controlling" the economy by fiscal policy, or the regulation of governmental spending, some economists hold that the government can regulate the economy by altering the growth of the money supply. According to these economists, when the economy is sluggish, the growth of the money supply should be increased; and when the economy is growing too rapidly, the rate of growth can be slowed by lowering the rate of increase in the money supply.

The difficulty with both these means of regulating the economy is, once again, that of the economic time lag. In 1967 and 1968, for example, growth in the money supply was high, reaching a peak in the latter year, whereas it was not until a year later, in 1969, that a corresponding peak was reached in industrial production. By then the government had become primarily interested in combating inflation. Accordingly, in 1969 the increase in the money supply was almost cut in half, but industrial production did not fall back sharply until 1970. It could be argued that the 1970 decline in industrial production occurred because the markets for a number of major industries were becoming supersaturated—most notably by aerospace.

We are thus confronted with two problems: first, that the behavior of the economy is subject to a great many variables, so that it is extremely difficult to determine how much of a given effect is attributable to a particular action by the government; and second, there is a lag of at least one year before the effect of that action can be observed.

Suppose, for example, that there is an effort to stimulate the economy of a region by undertaking a large construction program. The processes of planning, awaiting bids, design, site preparation, building, training staff, and getting an installation into operation inevitably mean a long delay. In the building of a regional power facility, fifteen years might elapse from the time the project was first conceived until it became effective. Government action to stimulate a regional economy might involve assembling and training a team of advisers on a new kind of farming, for example. Here, once again, time would be needed to put such a project into operation. As a higher and higher level of technical expertise and more sophisticated management techniques are employed, the total delay can be a matter of years.

There are, moreover, several mechanisms that operate against reversing a trend before it can get out of hand: short-term business fluctuations that tend to obscure underlying basic developments in the economy; the tendency of businessmen to err on the side of optimism during a long period of economic expansion, and thus to order more of everything than they need—experience having taught them that the penalties for expanding too slowly are greater than those for expanding too rapidly.

The long time lag before a needed decision is made on the question of air pollution and respiratory mortality is especially poignant. There may be a lag of anywhere from ten to twenty years between the time pollution becomes a danger and the time the effects show up in a rising incidence of lung cancer. Thus many people may have been condemned to die long before the need for a massive attempt at pollution abatement is recognized.

Perhaps the most serious time lag of all occurs in legislative action. The operation of the party system tends to slow down the passage of urgently needed laws. The wish of an opposition party to prevent a President from looking good in an election year, for example, can override all other considerations. At a time of acute strains on the social and economic system, when crises may occur without warning, the inability to respond speedily to unexpected developments has limited the effectiveness of our lawmakers.

The solution to this problem would seem to lie in appointed regulatory bodies, with both investigative and monitoring authority, and powers to take emergency action where the need is clear. For example, a government agency could be authorized to monitor the number of applications for building permits, and to deny issuing further permits as a check on overbuilding—thus preventing needless waste of building materials and manpower. A similar government agency might be empowered to oversee management decisions by airlines and other transportation companies, to forestall such things as bankruptcy because of low load factors.

Corporations will object to this as unwarranted interference by the government in purely corporate matters. When a corporation is very large, however, its decisions become a matter of public concern simply because of the large amounts of labor, capital, matter and energy involved. If corporate decision-making had a record of being generally right, such monitoring would not be called for. But its spectacular errors have now become a matter of public concern.

14

SCENARIOS OF THE FUTURE

By examining the way decisions and policies affect the nature of the future, we can decide what to do now in order to get the particular future we would like. It is extremely important to understand that a "decision" or "policy" may not be the result of conscious choice. Failure to take a certain action may be equivalent to deciding to take an opposite action. Thus, if we fail to limit growth in demand for energy, this has the equivalent effect to deciding to allow it to grow. Thus, we may distinguish between "explicit," or conscious decisions, and "implicit" or unconscious decisions.

Mankind has survived up to now without an elaborate science of "futurology." But the impact of human activity on the planet is today far greater than ever before, and is rising by a *factor of two* every decade. Formerly, if Spain or Greece cut down too many of their forests, or if Iraq allowed its irrigation system to deteriorate so that the soil became too saline, the effect, however severe, was still only local, and did not prevent the development of cultures elsewhere. However, we have now reached a stage in the evolution of civilization where humanity itself is a principal determinant of the future of the planet, everywhere on the planet. We influence the weather, globally, and the amount and character of the vegetation over large sections of continents. We are the principal determinant of the character of the ocean surface, and have a significant impact on the surface of remote glaciers. The moisture and chemical composition of the soil in large regions is man-determined, as is the acidity of rain in many places. If civilization now errs, and chills the planet with air pollution, or depletes the stock of a critically important resource or resources, or introduces a subtle, slow-acting poison into the biosphere, the effects will be pervasive and ubiquitous. It is unlikely that there will be significantly large undisturbed or unaffected refuges from which a new civilization could emerge if the dominant present civilization makes a serious blunder. Not only must we think seriously about possible future consequences of present actions, we must also think about the potential consequences of extrapolating various future trends. For example, if we have this much effect on air pollution when world total consumption of energy is the equivalent of seven billion tons of coal a year, what will it be like in the year 2000 when it is the equivalent of 45 billion tons a year? What will happen to life in the oceans when oil is being spilled on the surface at over six times the present rate?

A second argument for the importance of futurology is becoming important: the "irreversibility argument." Civilizations go through certain natural sequences of events during their evolution. It is extremely difficult to leapfrog stages in these series without external help. For example, a logical sequence in energy use by a developing civilization is to use first fuel wood, then coal in progressively deeper seams, then crude oil in progressively deeper and less accessible pools, then hydroelectric power, natural gas and nuclear fission and fusion in that order. The energy made available by the first items in the series allows development of the technology which then permits exploitation of the later energy sources in the series. For example, since it takes enormous inputs of energy even to do research on nuclear energy, it is difficult to see how a civilization could leapfrog use of coal, oil and gas, and go directly from fuel wood to nuclear fission. Likewise, in metallurgy: it would not be possible to contain the temperatures necessary to create a controlled fusion reaction in materials whose manufacture used no more heat than that obtained from fuel wood.

Civilization is now stripping the planet everywhere of all fuel and non-fuel minerals that can be obtained without the use of vastly sophisticated technology. Thus, if we err, as Egypt, Babylon, Greece, Rome, Persia, and all the others did, then the climb back to the present level of civilization will be enormously more difficult for any subsequent civilization than the climb to here was for us. In other words, humanity may have just one chance to attain a very high level of civilization on this planet, and we are involved in that chance right now. Thus, we should analyze our strategies, decisions, and policies carefully, to ensure that we are not cutting rungs off a ladder behind us, that others might want to climb up later, centuries or millennia from now. This type of thinking was irrelevant for previous civilizations; they simply lacked the power to cut rungs off the ladder behind them.

Since some possible scenarios for the future are intensely traumatic, they may provide the motive for an exercise in self-determination—all the more because the prospect of trauma does not lie generations ahead, but decades.

VARIABLES WITH IMPLICATIONS
FOR THE FUTURE

Since energy is the means by which other resources can be made available to us, the availability of energy is a critical determinant of the future. Even very low grade ores can be useful sources of minerals in short supply if we have enough energy for the extractive process. Also, the productivity per unit of cropland is determined by the input of fertilizer, which in turn depends on the availability of energy to extract, manufacture, transport and apply the fertilizer. For this reason, energy ultimately determines the "carrying capacity" of the planet for people: the more energy that goes into agriculture, in the form of fertilizer, farm machinery, and so on, the more people can be supported by an acre of land. If the energy input into agriculture is diminished, because one source of energy can not be replaced by another when the first source runs out, then the carrying capacity decreases. This inevitably would result in large numbers of deaths due to starvation if the area affected was large enough.

The following phenomena and decisions will play critical roles in determining if future energy supplies are adequate to meet growing demand.

Technological "fixes" or "breakthroughs" making possible energy-generating systems are necessary to replace the oil, gas, and cheap uranium oxide that will soon be gone. These new systems are also important to the underdeveloped countries, because they will allow their populations to reach a level of industrialization at which children will be perceived as an economic liability rather than an asset. Thus, having sufficient energy for the world is one means of controlling population growth.

Some readers may wonder why it is necessary to get energy systems breakthroughs "urgently." The reason is that it takes a very long time from the time a new discovery is made to the time when it is supplying energy to a nation on a significant scale. Thus, if we should run out of conventional energy sources in the period 2000 to 2010, we should be in the process of replacing them already. To illustrate the long time involved in converting a nation to new energy-generating systems, the following table shows the elapsed time from beginning of use of a new energy source to the time it was supplying ten per cent of U.S. national consumption.[1]

Admittedly, all these transitions took place under non-emergency conditions. However, even if a transition to fusion or solar power took place under a state of national emergency, it is difficult to see a conversion to any one of these as a supplier of 50 per cent of national energy requirements in less than 30 years. Thus, if we are to need supplementation of the national energy supply by solar or fusion power by 2000, we should be beginning installation now. If political leaders make speeches to the effect that new energy sources will be available by 2000 or 2010, the electorate will be able to evaluate the realism of such statements by keeping this time lag issue in mind.

An extremely important factor in determining the future state of energy supply and demand will be the intensity of the efforts to regulate demand. It is unlikely that we will be able to balance supply and demand by increasing supply indefinitely, so if there is to be any regulation, it will have to involve curtailment of demand. This can be done simply by making society more efficient with respect to energy use. To indicate how this could be done, more use of car pools, and staggering working hours so a quarter the number of commuter buses and trains could carry everyone to and from work, and be full most of the day would together reduce transportation systems energy requirements by a factor of three or four. Change in building design to minimize heat flow out in cold weather, and into buildings in hot weather would also cut down energy requirements to a surprising extent. The principal factor here is the extent to which the public will be willing to accept such changes, and legislative bodies will be responsive to the public will.

Rate of investment in research and development of new energy sources will have an immense effect on the shape of the future. If heavy investment doesn't begin soon enough, alternate energy sources will not be ready when they are needed. To this point, neither the public nor politicians seem very alarmed about this prospect. Whether or not both groups change their level of concern is crucial for the future.

As underdeveloped countries discover that they can improve their economic position by speculating in their energy stocks, rather than selling them soon for low prices, this will have a great effect on the availability of energy for the developed countries. Since the developed countries have already used up much of their own fossil fuel energy stocks, they are very dependent on supplies from the underdeveloped countries.

The future of energy availability will be strongly influenced by the extent to which capital flows from developed to underdeveloped countries. On the long term, such capital flow would have the effect of speeding industrialization and lessening the population growth rate. On the short term, the resultant industrialization would make the underdeveloped countries important competitors for the fossil fuel energy the developed countries are running out of.

The content of advertising about energy will have an important impact in shaping the future. Many oil and gas companies and electric utilities are now publishing advertising designed to inform the public about the energy crisis.

Minor changes in the content of this advertising may have rather large political consequences. Perhaps the most critical item is whether or not the energy supplying companies mention how limited the ultimate recoverable stocks of fossil fuel are, and how soon they will all be gone unless there are stringent attempts to conserve them. No advertising that I have seen has mentioned that theme. What the advertising states is that while there is an impending energy crisis, it could be alleviated by making more money available to the supplying corporations with which to explore still more energy. The effect of this increased exploration money would be to simply use up the world stocks even faster than they are now being used. A much more rational approach would be to increase the price of the fuels. This way, the suppliers would have more exploration money, but also, the public would be reminded by the price increases that fossil fuels were becoming a scarce resource. If all fossil fuel supplies were to be increased in price by a factor of three, we would quickly have a much more rationally run world, with modern, efficient, pleasant, and fast mass transit, for example, and much better designed homes and cities.

Along with population size, density, and growth rates, the variables to watch in measuring the energy situation are the total energy use per capita, nationally and worldwide, the price of gasoline, the use of crude oil, gas and coal, the cumulative use of crude oil, gas and coal, and the up-to-date estimates of reserves. (see p. 204)

The most important single variable shaping the future will be the birth rate of women fourteen to twenty-four years of age. This, in turn, will be influenced by the structure of the population, the unemployment rate, the wholesale price of beef and wheat, and the annual rate of change in the implicit price deflator, which measures inflation. The annual rate of increase in the price of basic food commodities will also be a means of projecting likely states of the future. The content of advertising will have an impact on birth rates. If it gives the impression that all is well with the world, and discounts evidence of increasing prices and other difficulties, this could elevate birth rates.

Death rates will also have an impact on population growth rates and sizes. Important variables to look out for will be epidemics due to infectious diseases, which are always more serious at higher than at lower densities, the effect of pollution on the respiratory disease death rate, the effect of pollution on crop growth, either directly, by poisoning plants, or indirectly, by making the climate deteriorate, and the life expectancy at birth.

A small number of variables are critical in determining if economic difficulties are on the way: the take-home pay of factory workers, in deflated dollars, the balance of trade, the rate of increase in retail sales in deflated dollars relative to the rate of increase in population, the automobile sales rate, the amount of consumer debt outstanding per capita in deflated dollars, and the gross national product per capita in deflated dollars. Housing starts and coal production are also important measures of future behavior of the economy, as are orders for machine tools. Many of these statistics are conveniently arrayed in the monthly government document "Business Conditions Digest" produced by the Bureau of Economic Analysis in the U.S. Department of Commerce. The other statistics can be calculated from the statistics listed in this publication, in most cases.

The effects of pollution are measured in such documents as the annual reports of the President's Council on Environmental Quality. Particularly important indices are concentration of oxidant (the gaseous pollutant most strongly influencing the urban death rate), and the global mean temperature depression, for predicting crop growth.

SOME PARTICULARLY CRITICAL TIMETABLES

Certain variables will be so important in shaping the future that we include timetables here, showing how they will change under specified assumptions, if no positive action is taken to change the timetable.

The first of these is the timetable for total world depletion of crude oil. The timetable shown in Table 1 has been prepared assuming that ultimate recoverable total amount is 2100 billion barrels,[2] and that production grows world-wide at 7.0 per cent per annum.

Three years of consumption are included beyond the date for depletion of world crude oil under this schedule, because many people assert that tar sands or oil shale could replace crude oil. They could, and the amount available could keep us going for exactly three years, given current estimates of the available quantity.[3]

This table can be used in several ways. The reader can see if we are on track by comparing these numbers with the annual statistical reports for the world which are published in many places, including the Statistical Abstract of the United States. This document is widely available on news-stands about January of each year, as the American Almanac, published by Grosset and Dunlap.

If some effective national or international steps are being taken to conserve fossil fuels, this will reveal itself by the items in the first column increasingly overestimating the actual data as they come in. On the other hand, if some policy or institutional practice, such as the content of advertising, is actually promoting the rapid depletion of fossil fuel resources, that would be evident because the statistics in the first column of the table would underestimate actual future use. The table thus serves as a check on the effectiveness of resource conservation policies.

Another important timetable for showing how the

TABLE 1   World depletion of crude oil

Year	Billions of barrels of crude oil produced	Cumulative use in billions of barrels	Amount left, in billions of barrels
1973	20.4	334	1765
1974	21.8	356	1743
1975	23.4	380	1719
1976	25.0	405	1694
1977	26.8	431	1668
1978	28.7	460	1639
1979	30.7	491	1608
1980	32.8	524	1576
1981	35.1	559	1540
1982	37.6	596	1503
1983	40.2	637	1463
1984	43.0	680	1420
1985	46.0	726	1374
1986	49.2	775	1324
1987	52.7	827	1272
1988	56.4	884	1215
1989	60.4	944	1155
1990	64.6	1009	1090
1991	69.1	1078 (half gone)	1022
1992	73.9	1152	948
1993	79.1	1231	868
1994	84.7	1316	783
1995	90.6	1406	693
1996	96.9	1503	596
1997	103.7	1607	492
1998	110.9	1718	382
1999	118.8	1837	263
2000	127.1	1964	136
2001	135.9	2100	0
2002	145.5	2245	
2003	155.7	2401	
2004	166.5	2567	

world is progressing into the future is the schedule of population increase. Measures to limit population growth

TABLE 2   Projected world population growth

Year	World population in billions	Year	World population in billions
1975	4.00	2011	8.79
1985	5.00	2012	8.97
2005	7.81	2013	9.15
2006	7.96	2014	9.33
2007	8.12	2015	9.52
2008	8.29	2016	9.71
2009	8.45	2017	9.91
2010	8.62	2018	10.11

worldwide are only successful if actual population growth is severely overestimated by the projection in Table 2.

Under this schedule, which assumes a 2 per cent per annum growth rate beyond 1985, it appears that even with extraordinarily rapid international installation of a fusion reactor energy supply system, most current problems would have become insurmountable, simply because of the press of human numbers. The amount of food available per person would be uncertain, and the cost of food would be astronomical. The table can be used to show if efforts to ameliorate those problems by control of population growth rate are proving successful.

Over the very short term, a factor which will have a large effect on the economy is inflation. A good measure of the impact of inflation on the average family is economic index number 752 used by the Bureau of Economic Analysis in the Department of Commerce, and published monthly in Business Conditions Digest. This index is the wholesale price index for farm products. Since this index shows a seasonal pattern of fluctuation, Table 3 only projects the value each year for June, assuming an 11 per cent per annum rise each year for the years beginning 1973.

TABLE 3   Wholesale price index for farm products

Year	Index	Year	Index
1967	100.	1974	149.9
1970	109.5	1975	166.4
1971	113.8	1976	184.7
1972	121.7	1977	205.1
1973	135.1	1978	227.6

If prices increase at a less rapid rate than indicated by this table, then inflation is at least partially controlled. If inflation increases as fast as, or faster than the projections in this table, then it will be a sufficiently serious problem to have implications for other sectors of the economy. For example, if inflation continues to increase this rapidly or faster, then it would probably have an impact on sales of durable goods, which could set in motion a vicious spiral of layoffs and increasing unemployment and still less purchase of durable goods.

FOUR ALTERNATE FUTURES

Given the preceding background, we now proceed to construct four scenarios, indicating different yet plausible paths by which we might move into the future from the present. The scenarios need certain attributes to be of some use. They should make some reference to quantitative measures indicating which state the world is in, at any point in time. They should specify dates of important discoveries or breakthroughs, so we know if such have occurred in time to be effective in preventing certain disasters from occurring. We need to specify early warning symptoms of good or bad situations developing. Finally, we must pinpoint critical mechanisms facilitating or blocking change in each of the unfolding scenarios.

Possible futures were generated by constructing logical

Will there be a continued high rate of population increase?	Will market saturation limit growth in G.N.P.?	Will depletion limit growth in resource use?	Will pollution be controlled?	Scenario number
			Yes	1
		Yes	No	2
	Yes	No	Yes	3
			No	4
	No	Yes	Yes	5
			No	6
Yes		No	Yes	7
			No	8
			Yes	9
	Yes	Yes	No	10
		No	Yes	11
No			No	12
	No	Yes	Yes	13
			No	14
		No	Yes	15
			No	16

branching tree diagrams to identify qualitative types of scenarios. The technique is illustrated in the figure on this page. Four questions were asked, to each of which the only permissible answer was yes or no. Thus, the number of qualitatively different scenarios that could be generated was $2 \times 2 \times 2 \times 2 = 16$. The four questions were:

(1) Will there be a continued high rate of population increase? (or will unemployment among the young and explosively growing inflation cause a great decline in the birth rate?)

(2) Will growth in the gross national product be limited by market saturation?

(3) Will growth in resource use be limited by depletion?

(4) Will pollution be controlled?

Of the sixteen possible scenarios, the four most interesting were selected because they have a high probability of occurrence. Of course, a spectrum of conditions could be imagined with respect to each of these questions, and an infinity of futures is possible. Our only objective here is to point out certain qualitative types of futures which are highly likely.

Quantitative details in connection with the scenarios are plausible in terms of trends that have occurred in the past, or trends showing up in the most recent available statistics, and newspaper and magazine reports. For example, while a spectacular drop in the U.S. birth rate may seem far-fetched, it was occurring in the spring of 1973. The scenarios which seem most likely to occur are numbers 7, 11, 13 and 15.

Scenario 7. The "trouble-free, technological fix" scenario. Three critical variables all continue to rise: the average gross product per capita in deflated dollars, both nationally and internationally, the resource consumption per capita, nationally and internationally, and national and global population. There are no major efforts over the next two decades to conserve resources, because all large institutions are increasingly characterized by boundless optimism about the future, and rigidity in decision-making. Resource use and population increase in accord with the preceding year-by-year schedules. However, happily, all the technical problems with the conversion of coal to gas and oil, and home heating and cooling by rooftop solar energy collectors and breeder reactors are solved by 1980, and a massive, emergency program is undertaken to install these energy systems nationally and internationally beginning in 1982. There is a high level of pollution control beginning in 1980, so that even with a level of energy use in 2000 about eight times that of 1970, on a worldwide average, there is no significant increase in death rates, or alteration of the weather.

Industrialization spreads and develops rapidly in many underdeveloped countries from 1980 onward, simply because of the money that can be made out of these countries as investments.

However, the cost of living becomes ferocious. By the year 2000, gasoline sells for $3.22 a gallon. Everyone hopes that with the conversion to a fusion-reactor based economy, prices will drop, but instead they rise still further. The prices of food, clothing and shelter are fantastic, and a college education is necessary to both the husband and wife if a young couple is to survive in the intense competition to get jobs. Having a child is perceived by all young couples as an unmitigated economic disaster. Scarcely anyone has a child before age twenty-eight, and only the wealthy have two children. This perception of children as an economic millstone blocking off upward economic mobility for young marrieds rapidly spreads internationally. Dropping birth rates spread like wildfire, and the world population, which had been 10 billion in 2018, has dropped to 8.97 billion by 2050, and 7.56 billion by 2100. Throughout the period, death rates per 1000 per year stabilize at 9.4, and birth rates stabilize at 6.0. This leads to a population drop of 3.4

per 1000 per year. Birth rates are less than a third the 1970 level of 18.2 per 1000 per year in the United States. Population keeps dropping for centuries, until finally it stabilizes in 2720 at 920 million people. If this scenario comes about, it implies that the ultimately stable state for the world is a population about equal to the world population of 1840, living in a society characterized by a high level of economic activity, and a high per capita level of resource consumption. It is highly unlikely that this scenario could come about, for several reasons. Pure luck followed by an organized effort by a galvanized society to install new energy technologies quickly in a desperate race against time is an initial assumption. It is unlikely that we would be that lucky or that organized. But further, it is difficult to see how mankind could get through the period 2000 to 2720 without major disasters. This would be a period with a very high population, and no gas, oil or coal, and society totally dependent on perfect functioning of very advanced technologies. No one experienced in such matters would have this degree of confidence in the perfect functioning of technologies. Brownouts and blackouts would be of enormous and unprecedented scale, and would cause immense hardship and death in winter in northern climates. Slight changes in weather or massive pest outbreaks covering whole continents would cause terrible famines. Also, it is difficult to see how we can get from here to the time of the "technological fix" without a variety of serious economic problems that impede population growth and resource use. Either market saturation in developed countries, or speculation in natural resources by underdeveloped countries leading to massive price increases for gasoline would prevent this scenario from becoming reality. It is too vulnerable to interruption at many key points, and must be regarded as optimistic, and quite unrealistic.

Scenario 11. Economic decline because of market saturation. Beginning in 1974, there is a sharply reduced rate of economic growth in the developed countries because of market saturation, and supersaturation of the labor market. This first shows up in the aerospace and electronics industries, and shortly thereafter in the automobile industry and housing construction industry. Unemployment rises significantly in November of 1973, and is 5.7 per cent of the labor force in February of 1974. It rises about a tenth of a per cent a month for the next three years. The rate of growth in earnings per share in American corporations flattens by the spring quarter of 1974, and there is a massive exodus of footloose capital to underdeveloped countries, in which market saturation is at least four decades in the future.

Unemployment reaches 25 per cent among teen-agers, and 15 per cent among people 20–24. A widespread view in these ages is, "We wouldn't dream of having any children. We can't possibly support ourselves, so how could we support them?". The populations of first the developed countries, then the underdeveloped countries start to decline. The British population begins declining in 1973, and that of the United States begins declining in 1979. Birth rates in the developed countries drop spectacularly to a third their 1970 level by 1985. Young people delay childbirth until they are twenty-eight, and the two-child family is a rarity. This decline in the birth rates spreads very rapidly throughout the world. Only six million cars are manufactured in the United States in 1975, and only five million of those are sold.

The population ultimately stabilizes at one billion persons worldwide. New energy sources are gradually discovered, but are not critically necessary because the use of solar energy, fuel wood, and a wide variety of measures for economizing on use of energy cuts down demand for other forms of energy.

An interesting feature of this scenario is the behavior of governments everywhere during the 1974–76 crisis period. A variety of measures to stimulate the economy are tried, but they all make matters worse. The amount of money in circulation is increased by 15 per cent during 1974, and several massive contracts are awarded for the construction of space stations, space shuttles, missions to Mars, and supersonic jet transports, but the result is enormous increase in government debt and runaway inflation. There is a great deal of confusion in government, and inability to galvanize and organize to deal with the problems, because much of the business media insists that the situation is merely a temporary downturn in the business cycle. Few agencies are in existence within the government which have the techniques or the point of view required to analyze the economic phenomena involved, or make recommendations for effective actions to ameliorate and rectify the situation.

Gradually, the situation is corrected, by increasing recourse to fiscal as opposed to monetary policy for management of the economy by the government. Money is directed toward urban renewal, mass transit, and an enormous expansion of education, health, research, culture, and various other service industries. The economy recovers gradually, but the lesson which has been learned spreads internationally. Population sizes decline everywhere, and there is great concern about efficiency in the use of matter and energy. Market saturation is carefully avoided.

Scenario 13. Population declines because of resource depletion. Crude oil consumption worldwide increases in accord with the schedule in the table, and many government and business leaders insist that there is no need to worry about resource depletion. "Fusion and breeder reactors will solve everything." Then we run out of crude oil, on schedule, and any replacements are inadequate, and the technologies can not be spread around rapidly enough.

Prices of everything, particularly food, skyrocket, and the birth rate drops precipitously everywhere. The world population ultimately stabilizes at one billion. The early warning sign for this scenario is that by 1980, no new technology that would have to be ready then to be in place everywhere by 2000 is available. Statements by scientists about the timing of the arrival of fusion energy continue to be of the form, "Don't worry, it will be available in thirty years. We just have a few bugs to iron out." The statement has taken this same form for the last thirty years, and the estimated time to project completion still has not shrunk one year.

Scenario 15. Sharply reduced birth rate because cost of living becomes exorbitant. China and Russia buy up increasingly large amounts of food from United States and other net exporters of food, and prices of food rise at 20 per cent per year. Everyone is in debt to the limit of their credit. Retail sales begin to fall off, automobile sales particularly fall off, and birth rates continue to decline sharply. By 1985 the U.S. population is declining. But high cost of living pinches hard in all countries, and this leads to birth rate declines and reduction in sales in the developed countries, and massive famine in the underdeveloped countries. World population declines at a third of a per cent each year, beginning in 1985, when population is five billion. By 2000, the population is 4.76 billion, and by 2100 it has declined to 3.4 billion. By 2500 it has declined to 897 million. Ultimately, the world population stabilizes at about a quarter of the 1973 level.

Gradually, the world economy stabilizes in a pattern

characterized by low population densities, a high proportion of the labor force employed in a very wide variety of service occupations, and a tremendous concern with efficiency in the use of matter and energy.

All four of these scenarios have several things in common. They assume an ultimate population of the world far lower than what we have now, and no international conflict. I assume no wars because nuclear holocaust will be unthinkable for all countries, and in any case, the food stores of many underdeveloped countries will be too low to support war. The population will be literally too weak.

No miracles are assumed. It is possible in principle that someone should discover a way of obtaining infinite amounts of energy from gravitational fields or time reversal, but extremely unlikely. Breakthroughs in our understanding as important as those of Newton or Einstein occur only very rarely, in the minds of extremely unusual people. We can not count on such developments occurring on a fixed schedule as needed.

It is highly likely that our future will be some combination of scenarios 11 and 15. The early warning flags are up already: escalating food prices, market saturation in certain industries, unemployment among the young, and dropping birth rates.

Such a scenario would have a high probability of persisting once it develops. The world would have a smaller population, with a much lower rate of resource depletion and pollution. There would be less people to use resources and pollute, but also a lower rate of pollution and depletion per capita. There would be a very high level of employment, and indeed, society would be designed so as to ensure this. Because of the high level of research activity, solutions for the problems of society would typically be available before any problem became serious.

This high level of research would be like social insurance: protection to minimize the chance that society could get into difficulties because of unforeseen causes.

The world would be very attractive, as it was in the middle age of cathedrals, for example, because there would not be a systematic avoidance of labor-intensive activities, as at present.

Since resources would be depleted slowly, there would be plenty of time for the research to find the best ways of recycling resources, or to find ways of developing alternate sources. The overall tempo of change would be slower than at present, because population growth and the rate of resource depletion and pollution would be slower. Thus, there would be more time to deal with problems. The present situation, in which we are always rushing headlong into catastrophes would be done away with. There would be more time to relax. The wise use of leisure time would become very important. There would be many satisfying ways to spend leisure. Good music would be more widely available. Art and architecture would be more attractive and there would be more to see, that was worth seeing. Television, the radio, and motion pictures would provide more interesting and higher quality entertainment.

It is unlikely that this scenario would be disturbed by major problems. It would involve a great deal of social and economic self-regulation, or balancing mechanisms. Life would be pleasant and even.

THE ROLE OF GOVERNMENT DEBT

Some readers may wonder why it is not possible to solve all problems simply by increasing government spending, without regard to the magnitude of the resultant debt. Is there any ultimate limit to the size of the federal debt which can be carried? The following scenario indicates what the consequences would be if this solution was employed.

Because of a series of years in which government income was lower than expected, due to unemployment, and expenses were greater than expected, due to welfare costs resulting from the unemployment, the federal debt burden becomes enormous. Gradually, the ability of the government to deal with major new crises is limited by restrictions imposed on the growth rate of the national debt. Finally, problems arise which are so great that the government simply cannot deal with them. Large numbers of very large corporations with massive staffs go bankrupt and must fire their staffs. The unemployment problem becomes still more serious. The problems of Lockheed in 1971 were probably our first hint of this type of problem occurring on a large scale. The government could deal with a few problems of this type, and on this scale, but the funds required to deal with a large number of them simultaneously would simply not be available because of limitations on the rate of growth of the national debt.

A problem of depletion of capital in the United States cannot simply be dealt with by decreasing the value of money (through deficit spending or increasing the money supply), because this creates international economic difficulties which are difficult or impossible to solve. Ultimately, money must have some commonly agreed on value, and people must believe that that value will remain relatively constant, or the entire monetary system collapses. No one will invest money anywhere if it is going to lose half its value in a year, for example.

The alternate futures sketched here are only four of many which could be imagined. There are some common features to all of them that some people will find odd. For example, they all assume that at some time in the next fifty years, there will be such a dramatic decline in birth rates that population sizes will decline. However, it should be noted that there is rapidly growing concern in many countries about population size. Singapore, for example, has recently decided on population control, as has the Association of Bay Area Governments in the San Francisco Bay area.

One of the most unusual features of all these scenarios is that they do not visualize government as being an important change agent. Changes are assumed to occur because of market mechanisms (high prices operating to discourage reproduction as well as retail purchases), and changes in popular thinking.

But is this view of government as largely ineffective and irrelevant a necessary view? In the next, and last chapter we consider techniques for making government a change agent in helping to produce a more rational society.

15

POLITICAL REMEDIES

In retrospect, it appears that the period since the Second World War has been a consumer's golden age for many Americans—the ultimate result of an exploitation ethic deeply rooted in American history. Kenneth Boulding refers to this ethic as the "cowboy economy"—Go West, young man, and make your fortune! The frontier promised boundless resources to exploit without regard to efficiency of use. One could always move on when local resources were used up or pollution became unbearable.

Massive inefficiency in the use of resources has continued into the present, long after the frontier expired. As a result of cheap energy and agricultural land, the cost of basic necessities (food, housing, and clothing) has taken a progressively smaller proportion of the typical family income. For example, the cost of food only increased 2.02 per cent per annum during the period between 1950 and 1969 while median family income increased at an average rate of 5.65 per cent per annum during the same period.

As the cost of basic necessities goes down, money available for the purchase of luxuries goes up, and lifestyles consequently become more resplendent. Now we would indeed be fools to complain about a substantial improvement in our lifestyles, *if there were no hidden costs!* Unfortunately, the costs are there. They have simply been deferred. Cheap agricultural land translates into urban sprawl. Cheap gas translates into freeway congestion and air pollution. Cheap food translates into an environment overloaded with pesticides and fertilizers.

The important point is that the situation is changing amazingly fast. In recent months food costs have begun to rise explosively. In the very near future, gasoline prices may double, and other fuels may be rationed. There is talk of electrical brown-outs. In short, the hidden costs of resource exploitation are being exposed. It has suddenly become reasonable to think about mobilizing a political constituency interested in conservation of energy, land, and other resources. Political action strategies must be implemented to reduce hidden environmental costs and to make more efficient use of resources in America.

THE POLITICAL GOAL

Numerous grave problems for American society have been discussed in previous chapters: market saturation, pollution, inflation, manpower glut, and international monetary turbulence. Ultimately, the cause of all of these problems can be linked to the continued inefficient use of energy and other vital resources. Market saturation, for example, involves inefficient uses of energy, minerals, human time, and money. Efficiency is sacrificed to increased production. When energy and minerals are squandered, the end product is inflation. A manpower glut occurs when too high a proportion of the labor force attempts to earn a living in activities related to the production of goods. It is much easier to saturate the job market in production activities than it is to saturate the service sector of the economy.

The demand for nonphysical goods, both services and leisure activities, can grow to meet increasing supplies without harmful side-effects to the environment. International monetary turbulence stems primarily from an imbalance in trade. Trade imbalance could be largely avoided if we imported raw materials at a lower rate. Clearly, this would be accomplished if we made more efficient use of all raw materials.

From the above, we are forced to conclude that there is only one supremely important goal for American society at present: efficiency in the use of energy and other resources. If this goal is not attained, a whole panoply of clearly disastrous consequences will follow.

POLITICAL STRATEGIES

The efficiency of resource use is affected by two interrelated forces of market control: consumer demand and resource supply. Both high consumer demands and abundant resource supplies stimulate a tendency to squander resources. When consumer demand is strong, the high overhead of inefficient resource use can be readily absorbed. And, of course, when resources are abundant, there is no incentive to use them efficiently.

Political strategies can counteract the forces that work against efficient resource use. The government must work to decrease the level of consumer demand and at the same time alert the public that the U.S. no longer enjoys an abundant supply of virtually everything.

To control consumer demand, there must be a national commitment to achieve a stable population size in the United States as soon as possible. This is of paramount importance because there is a direct link between the size and growth of a population and demand for resources. The overall demand for any resource is its demand per capita multiplied by the population size. Demand per capita, in turn, is closely tied to personal living standards. As those standards rise, so does the use of resources. Rather than limiting living standards (which can be reduced only to the point where all people subsist at some minimal standard of living) it is preferable to limit population size.

A variety of specific measures should be employed by the government to minimize the birth rate. Legislative action could include the imposition of tax penalties for excessive numbers of children, and the denial of federal grant support to any hospital which refuses to perform abortion on demand. Federal funds could be used to disseminate information (concerning birth control and the actual cost of raising children) and to subsidize birth control-abortion-sterilization clinics where the services would be provided at cost.

A leveling off of the population growth rate will lead to reduction in the demand for energy and resources which will lead to a short-term surplus in the work force. To relieve this temporary excess, we must encourage the rapid development of service industries. Expansion of service industries will not only provide needed jobs, it will allow for economic growth in a non-resource-intensive sector of the market.

There are a number of specific tactics available to encourage the rapid development of service industries. These may include provision of low interest loans to small service oriented businessmen, partial government subsidy of technical schools for service personnel such as paramedicals and all types of repairmen, and the increased use of the armed forces as a training ground for service industry workers. A major national effort to build modern mass transit systems, viable recycling centers, and innovative housing for the poor could also take up any slack in the demand for labor.

Agriculture, the creation of new resources, can be considered a part of the non-resource-intensive market sector. American farmers have been selling out to agribusiness or development industries at an incredible rate. Although family size farms are generally more efficient in terms of energy/output and pollution/output ratios, the average farm size has increased dramatically in the last twenty years, especially in California. Tactics similar to those that encourage development in service industries are needed to put family farmers and family farm cooperatives back on the land.

Turning to the effect of resource supply on efficiency of use, it becomes abundantly obvious that the United States currently has no consistent energy policy. One of the major difficulties is that a large number of governmental agencies make policy decisions that have an impact on the use of energy. To illustrate, we have recently seen the Phase IV Price Commission attempt to prevent an increase in the retail price of gasoline. This implies that gasoline is available in such superabundance that there is no need to let the price float upwards in the interest of encouraging more efficient use of fuel. On the other hand, the President has led us to believe that petroleum is in such short supply that a mandatory national allocation scheme for fuel oil is necessary, and that the Navy's strategic petroleum reserves at Elkhorn, Nevada must be confiscated.

These two policies are distinctly inconsistent. If petroleum is in such short supply that we must have a mandatory allocation program for fuel oil or that we must confiscate the Navy's reserves, the Phase IV Price Commission should be allowing gasoline prices to rapidly increase. If, on the other hand, the Price Commission is right and gasoline is really available in super abundance as far into the future as we can project, then there is no need to have a mandatory fuel allocation program or to confiscate the strategic oil reserves.

These inconsistencies in energy policy are extremely significant. They are a vivid demonstration of the fragmented nature of energy policy decision-making throughout government. To correct this appalling situation, a national energy agency is urgently needed. Such an agency is needed at the federal level in order to ensure that it has the resources required for an overview of the entire energy situation, plus the power to make policy decisions concerning all aspects of energy supply, demand, and use.

The proposed agency should have the responsibility and authority to issue grants and contracts to promote new forms of energy production. The use of fusion, solar, geothermal, wind, and tidal power sources must be explored. The agency should also have control over current sources of energy supply through the regulation of power plant siting, and through the accurate projection of U.S. reserves of coal, oil, gas, and uranium available for future use.

The national energy agency should also be responsible for the accurate projection of the need for these energy sources. Private utility companies engaged in the selling of energy for profit cannot be expected to produce an unbiased projection of future energy demands. Until recently, a number of private utility companies in California were actively advertising the purchase of power-consumptive home appliances while at the same time projecting the specter of electrical brown-outs if new power plants were not constructed.

The new agency should also promote innovative methods of energy conservation, including setting requirements for efficient design of products, buildings, and transportation systems. It should include the setting of rate schedules to encourage the efficient use of all forms of energy (large energy consumers are currently given a discount), and the initiation of such consumer deterrents as a "horsepower tax" to discourage the purchase of large, gas-consumptive cars, in favor of smaller vehicles.

Another area in which this agency should be active is in the creation of public sensitivity to the energy problem. Cities and corporations should be encouraged to analyze both the benefits and costs of future growth, not only in terms of money, but also in terms of energy and other resource requirements. This could be partially accomplished by providing grants or contracts for the development of package programs designed to identify and evaluate the energy demands of projected growth.

Regional land-use planning should be promoted and coordinated by the agency to further sensitize the public to the complexity of the energy issue. Local land-use planning agencies will rarely concern themselves with the fact that if too much prime agricultural land is taken out of production to put to other use, fertilizer and other energy inputs into the remaining agricultural land will have to be increased in order to maintain agricultural productivity.

The recognition and correction of positive feedback loops at work in the aggravation of environmental problems is a final and rather novel approach that might be used by the new energy agency to improve the overall energy problem.

A positive feedback loop occurs when a given phenomenon exacerbates itself, as for example, when freeway construction stimulates the construction of more freeways, which in turn encourages more freeway driving, which results in still more freeway congestion.

Such positive feedback loops should be replaced with negative feedback systems which by definition are self-regulating. In the above example, for instance, freeway congestion should be countered by the construction of mass transit facilities which would in turn reduce freeway congestion.

Recognition of run-away positive feedback loops in environmental problems and their replacement by self-regulating negative feedback loops could be encouraged by the new energy agency through grant awards to study specific problems.

It is particularly important that transportation be handled on the basis of negative feedback loops because transportation uses such a large proportion of all the energy flowing through society. Appendix I amplifies the above example of positive and negative feedback loops in a transportation system. Appendix II gives an example illustrating the methodology to optimize the use of energy for transportation in a metropolitan area.

THE SOURCES OF POLITICAL POWER

Recently a bill was passed by the California legislature to set up a state agency empowered to do much of what was suggested above for a national energy agency. A great deal of effort was expended by a number of enlightened parties to draft and carry this legislation. Unfortunately the bill was vetoed by Governor Reagan.

It is instructive to see who supported the bill and who opposed it. The proponents of the bill were a diverse group including labor, environmentalists, public utilities, and a number of large corporations such as Dow Chemical and Hewlett-Packard. The opponents of the bill were the oil companies (whose opposition was relatively weak), private utilities, and the Public Utilities Commission (PUC). A question we might ask is "Why was the vetoed bill given such diverse support?"

Three points about the defeat of this bill are worthy of attention. First, the successful pressure by private utilities occurred because of their concern that energy policy decision-making remain fragmented throughout government. The private utilities are so concerned that this fragmentation be maintained, that they are prepared to go to great effort and expense. We call this the balkanization tactic.

It is expressly because of past balkanization tactics that energy decision-making is spread throughout a very large number of federal and state agencies. The same tactic applies in decision-making about land-use planning. This great division of responsibility and authority ensures that no single agency has a sufficient overview to have the knowledge or power for rational and comprehensive energy and land-use decision-making. Consequently, there is nothing within government that represents a serious threat to the decision-making that is taking place in the board rooms of large private gas, oil, or electric companies.

The second important point to notice is that regulatory agencies, in this case represented by the PUC, are frequently co-opted by the private interests they are supposed to regulate. As a result, private interests are served rather than the legislatively intended public interest. Perhaps this is also the fault of the enabling legislation.

The third important point illustrated by the defeat of California's energy bill is that private interest groups can almost always gather more money to push their interests in a legislative body than is the case for public interest groups. This is particularly true because private interests can pass on such "business expenses" to the consumer. Consequently a much larger amount of money can be translated by private interest groups into information, which is the critical resource affecting legislation. Efficiently and abundantly supplied information advances vested interests before legislative bodies. Legislative development by corporate interests will only be broken up when the public interest can be represented with the same resources that are now available to private groups.

The present coalition between private interests and government can only be changed if a series of reforms are carried out. Government officials must become politically accountable for their decisions. Public and private advocacy must be placed on equal footing. And finally, citizens' participation in government must be promoted.

POLITICAL ACCOUNTABILITY STRATEGIES

Political accountability is governmental responsiveness to the public interest. Failure is evidenced by the fact that legislative and administrative branches of government usually succumb to varying degrees of control by vested interests. The degree to which government is controlled, and the degree to which it is accountable to the public can be seen on a continuum.

At one end of the continuum are government agencies that are regulated by the industries or organizations they were charged to regulate. Here also are "bought" legislators. At the other end of the continuum are a handful of free-thinking legislators and public agencies who continually renew their efforts to uphold and defend the public's interest in the work they do. In this camp are those who believe that when persons assume the public's trust they should consider themselves public property and act in the public's view. Of course, most politicians play towards both ends at the same time.

Relationships between legislators and lobbyists operate to minimize the government's responsiveness to the public. As an environmental lobbyist once said of California legislators—"It doesn't surprise me that they can be bought off, but rather, that they can be bought off so cheaply." We are talking about more than campaign donations and borrowed cars, we are talking about lobbyist-prepared legislative proposals. There is a need for greater regulation of lobbyists. They should only be allowed to provide information. Financial and material favors should be discontinued, the office budgets of lobbyists should, nonetheless, be made available to the public.

As Watergate has so vividly demonstrated, it is essential to fund political campaigns from tax revenues, not private donations. There are at least two ways to keep the cost down: (1) strict limits can be imposed on campaign spending, and (2) the government can require that campaigns be partially subsidized by media and transportation. The Federal Communications Commission could require that a specified amount of media time be devoted free to each candidate, and the Civil Aeronautics Board could require that a specified amount of free travel time be given to each candidate by commercially scheduled airlines.

Political accountability should further be encouraged by a systematic effort to undermine political secrecy. All congressional committee meetings should be open, with votes tallied, and the minutes made readily available to the public. Citizen action groups should be encouraged to make grading sheets of government figures. A few organizations now publish such grading sheets, but this practice should become much more widespread.

An intense effort should be made to create an informed electorate. The government should sponsor TV specials presenting in-depth analyses of particularly important issues. The government, rather than privately supported television networks, should be providing leadership in the presentation of specials that explore aspects of current issues such as the supply and demand for energy, land, food, or other limiting resources.

All government officials (legislative, executive, administrative, and judicial) should be subject to tight conflict of interest regulations. For instance, no one from a private industry should be allowed to hold a position in a government regulatory agency associated with that industry. No officials should hold stocks or other financial interests in businesses which in any way may influence their public decisions.

A statement of the financial and real holdings of all government officials should be available to the public, and it should be made available in a standardized fashion to avoid confusion. Accounts of campaign spending and contributions should also be standardized. They should be open for public inspection, published, and up to date, even during the campaign season.

ADVOCACY AND CITIZEN PARTICIPATION IN GOVERNMENT

Some citizens may choose to participate in government by becoming members of advisory boards, or even by running for political office. This type of activity is usually referred to as citizen participation in government. Others will prefer to work outside the system, remaining autonomous and free from possible prejudice. The label for this group is usually "political activists." Often one person wears both hats. Both sectors are usually represented at important public hearings and legislative committee meetings. Coalitions of individuals or groups representing a wide range of interests are also becoming increasingly effective.

Political activists in the environmental movement come in all shapes and forms. Some readers may remember the

Fox, who thrilled local conservationists in Chicago with his daring guerrilla war against polluters. Among other pranks, one night the Fox managed to cap a smokestack. The next morning the noxious gases polluted the factory rather than the air outside. Discussed below are the more conventional reform tactics used by activists.

A person with enormous energy and talent can do an incredible amount of reform singlehandedly: Ralph Nader is the notable national example. Very few people, however, are psychologically or physiologically capable of working over 100 hours a week for more than seven years without a vacation. Nader—"Could you take a vacation if your house was on fire?"

Citizen action groups subdivide and organize reform and research efforts. They usually go on to serve other purposes as well. Information gathering and dissemination is a valuable service provided by most of the nationally based environmental groups. Besides news collecting, activity coordinating, and administrative personnel, the staffs of these groups often include Congressional lobbyists, resource specialists, and attorneys who seek redress in the courts. With paid staffs the donated time and money (dues) of members can be fully utilized. Few of these staffs could operate without volunteer assistance and the financial backing provided by membership dues. And, conversely, it would be nearly impossible for national organizations such as these to operate without paid staffs.

Many citizen action groups have chapter affiliations. This enables a group to be effective at state, local, and regional levels as well as nationally. Members can be more directly involved. Groups such as the Audubon Society and the Sierra Club also sponsor pleasure and recreational activities; often these are organized at the chapter level.

Resources for the Future, Sierra Club, League of Women Voters, and Friends of the Earth all have highly respected publications. The Committee for Environmental Information publishes a magazine entitled *Environment*. The Environmental Defense Fund, the Sierra Club Legal Defense Fund, and the Natural Resources Defense Council lead the list of organizations who pursue national or local reforms in court, that is, through law suits in the public interest. Providing Congressmen with information is a cadre of lobbyists from a large number of activist groups. Unlike vested interests, they provide only information. Common Cause and the League of Women Voters have expansive national programs which include environmental degradation in their priority issue categories. Friends of the Earth is a progressive young group with considerable international affiliations. Appendix III offers a select list of nationally active groups.

We believe that citizens will also become more directly involved *within* their government. Many people are becoming activated locally by the results of rapid and continued growth, and the lack of previous planning. The energy crisis, Watergate, food prices, and inflation all act as catalysts to encourage citizen action and participation at all levels of government.

Public participation in government is not new.

> I know no safe depository of the ultimate powers of the society but the people themselves; and if we think them not enlightened enough to exercise their control with a wholesome discretion, the remedy is not to take it from them, but to inform their discretion.
>
> Thomas Jefferson, September 28, 1820
> in a letter to William Jarvis

Citizen participation has been extensive in social areas such as health, education, poverty programs, urban renewal, and social planning. Public concern in these areas ran high for several decades, as demonstrated by the civil rights movement in the early and mid-sixties. A large faction of citizen involvement in government today is concerned with environmental issues.

The rapid expansion of activities by environmentalists began only about four years ago, and yet the movement is maturing. Purely environmental goals are modified and compromised by the processes that integrate them with other public goals. Local planning is a synthesis of social and environmental problem solving. An activist's desire to preserve social and environmental values intimately ties him to involvement with the planning phase of government.

Many policy and decision makers are now, or have been, in the process of developing a means for citizens to participate in government, especially in those areas where citizens are pressing strongest for the opportunity. In most cases this consists of newsletter services and public hearings. Currently, many government officials and citizens are together trying to discover the most effective means of utilizing citizen's efforts. It is evident to all those concerned that there are unaccommodated informational needs.

If citizens continue to press for greater opportunity to direct the goals and decision making in their government, then two things will have to happen to begin to accommodate citizen demands. First, the citizen must be supplied with information. For example, he will need to know the specifics about a policy decision, and when it will be made, and then he will need to know when, where, and how he can participate. Second, an increasingly more efficient means of using the input of citizens must be built into government. Institutionalized citizen advisory boards are asking for more than token participation.

Most government administrators, and in fact, most public interest activists, don't realize that failures in systems designed to incorporate citizen's participation result even when the goals of informing and consultation are more or less realized. Relationships between government and citizens are progressively more mature as partnerships emerge, as power is delegated, and as citizens gain control.

Coalitions of interests are emerging today, as they have in the past,* to meet problems of significant social interest with impact. Coalitions of a wide variety of groups with certain interests in common can create stronger mandates for social or environmental changes. The Sierra Club supported Shell Oil Company workers in their strike for safer working conditions. The California Program Action Committee of Common Cause, a coalition of interests in its own right, joined other activist groups in California in an effort to qualify the Open Government Initiative for the 1974 California ballot.

The potential for effective collective action grows with each significant advance made in the public's interest. When a proposal truly benefits the greatest need for the greatest number, more individuals, and hence, more organizations, will support it. We offer two quick examples from the thousands that exist. When airlines go to the Civil Aeronautics Board for price increases for Hawaii—mainland fares—environmental groups should support them. The Islands are suffering dire environmental and economic setbacks from an overextended tourist trade. When farmworkers strike for decent labor conditions, all working

*The political coalition (and constituency) built by Franklin D. Roosevelt illustrates the cyclic nature of government processes. Public interest groups are rarely strong enough individually to get legislation on the books. Roosevelt's response to this problem was perhaps the most sophisticated effort in this century. Roosevelt welded together a heterogeneous coalition of urban blacks, southeastern whites, blue-collar workers, university intellectuals, ethnic minorities, conservatives, leftist radicals, and middle class liberals. These people were united in their efforts to overcome the problems of unemployment and the other side-effects of the slow-down in the economy following the Depression.

people should support them, not only for humanitarian purposes, but also for environmental reasons. Large land owners get richer and bigger at the expense of the workers and at the expense of the health of the land. If they had to pay to provide decent working conditions the labor subsidy they get would be reduced, and the lower profit margin would limit their ability to expand.

A STEADY-STATE ECONOMY

If the goal of efficient energy and resource use is met through the successful implementation of the political action strategies discussed in this chapter, where does this lead? What will be the consequences for our current system of economics and our current lifestyle?

We foresee the necessity to modify our growth-oriented system of economics to a steady-state system. "Steady-state" is not meant to imply either static or stagnant economic conditions; nor should it conjure up visions of the U.S. economy brought to a standstill. Rather, a steady-state economy is based on the maintenance of a constant stock of physical goods and a constant stock of people (population). These stocks are sustained by continual inflow in the form of production of goods and births, balanced by continual outflow in the form of consumption of goods and deaths.

The rate of throughput (inflow and outflow) of these constant stocks of goods and people can be manipulated within natural limits. The goal of efficient use of energy and other resources suggests that we attempt to minimize the throughput of goods. Low rates of throughput imply, and often require, high life-expectancies. For people this is clearly desirable. With respect to goods, high life-expectancy is equivalent to high durability.

The foregoing is an elementary description of the mechanics of the steady-state economy. There are several excellent books available for those who would like to go into greater detail. A particularly readable book is *Toward a Steady-State Economy*, edited by Herman E. Daly. Daly has assembled a choice selection of papers considering the social, ethical, and fiscal ramifications of the steady-state economy.

The first critical step in the transition to a steady-state economy is the elimination of inflationary growth. Inflationary growth is one of the most insidious and vicious positive feed-back loops operating in our society. Who hasn't heard of spirals of inflation? It serves as a malignant catalyst to the processes of resource depletion and pollution. Inflation also deprives people of their savings, makes life difficult for people on fixed incomes, and worse.

One straight-forward method of eliminating inflation would be the institution of a permanent national wage and price freeze. As demonstrated in the Nixon administration, wage and price freezes have met with strong opposition from diverse sources, labor as well as business. Much of this opposition may be countered. For example, when labor spokesmen criticize freeze measures limiting earning potential, they should be reminded that goods and services will also have price limits. A wage and price adjustment board should resolve the inequities that exist as the freeze goes into effect. Incomplete contract negotiations and the like would require resolution by such a body.

As we approach the steady-state economy, the strategy of investors will necessarily change. (Games for Decision Makers, pg. 237, is included to help readers make this change to steady-state thinking.) The average price of stocks may go down as paper profits, based on the anticipated high rates of company growth, dissolve. However, there is no basis for the assumption that the value of the stock market will necessarily plummet. The underlying value of a given stock to an investor will remain constant, owing to the opportunity for increased dividends. That is, if companies are no longer forced to be heavily growth-oriented to survive, a higher percentage of the earnings could be dispersed to the shareholders rather than plowed back into the otherwise ever-increasing expansion of the company. Investors will not require a 7 to 10 per cent increase in the value of their capital outlay to keep up with inflation. Stock investments will represent the purchase of a share in the real profits of a company rather than a share of their potential astronomic growth. If inflation were eliminated, one of the principal forces for uncontrolled growth in our economy would disappear. More time would be available to achieve the goal of efficient use of energy and resources.

It should be emphasized that funding the conservation proposals we have been advocating will not necessarily place a greater burden on the taxpayer. Let's take two examples. First, many of the suggestions aimed at increasing political accountability are relatively cost free. Instituting stricter limitations on lobbyists has no cost. Funding political campaigns from tax revenues and from media and transportation subsidies may even reduce overall expenses. Second, while expense will be involved in the institution of a national energy agency, it should be recognized that the ultimate aim is increased energy and resource use efficiency. By definition, this will lead to a reduction in real costs.

Many people feel that new agencies add more levels to an already top-heavy bureaucracy. I agree that it is necessary to streamline the mechanics of our government. Institution of a new national energy agency possessing full authority and responsibility for all energy use and energy policy decisions can only occur if agencies with piece-meal responsibilities are required to give them up. Government would be reshuffled, duplicated efforts would be eliminated, and there would probably be less government, not more.

GAMES FOR DECISION MAKERS

This book suggests that much planning in the past has been defective, because of excessive optimism about future growth trends. It is of great importance to all organizations, particularly corporations, that managers be able to project future trends realistically. If future level of demand for a product or service is overestimated, then net profit will be depressed by unnecessarily high overhead. If, on the other hand, demand is underestimated, then opportunity for services or profits will have been missed.

The reader may want to sharpen up his skill in projecting future trends. Accordingly, I include some games designed to provide some insight into the causal systems operating on typical, or important sectors of the economy. (Is the market approaching saturation? Why? When will the market be so saturated that growth in demand will stop? Are there other limits to growth, such as inflation or depletion? What are the most important factors ultimately regulating each sector of the economy?)

After having entered your projections in the allocated columns, put the pages away where you can check your forecasts yearly.

GAME 1

Project future trends in domestic airline business. Do you think the airlines ordered too much equipment because of overly optimistic forecasts?

(Tabulated data is the billions of revenue passenger seat miles flown by U.S. scheduled air carriers on domestic flights. Statistics from *Statistical Abstract of the United States*).

Year	Airlines and aerospace industry projections (assumes 10 per cent per annum growth rate)	If airlines and aerospace industry have overestimated, growth may be as listed below (assumes 1.73 per cent per annum growth rate)	Enter your projections below	Enter the date of your projections in this column	Enter the reasons for making your assumptions in this column	Enter actual data as they become available
1969	102.7	102.7	102.7			102.7
1970	104.1	104.1	104.1			104.1
1971	106.3	106.3	106.3			106.3
1972	116.9	108.1				
1973	128.6	110.0				
1974	141.5	111.9				
1975	155.6	113.8				

Which projection was closest to what actually happened? What does this tell you about the assumptions underlying the projections? How close to the facts were your projections? What does this tell you about your own assumptions? Would you make the same assumptions now? Were your past assumptions objective, or were they influenced by propaganda about exponential growth? Could such propaganda be affecting your day-to-day business decisions adversely?

GAME 2

Project the impact that gasoline prices will have on the amount of travel by bus.

| Year | Average annual cost of gasoline as a percentage of median annual income of male employed persons[a] | | Percentage of all vehicle miles travelled by buses[b] | |
	Your projections	Data as available	Your projections	Data as available
1968	2.1	2.1	.6	.6
1970	2.1	2.1	.6	.6
1973				
1974				
1975				
1976				
1977				
1978				
1979				

[a] 1971 SAUS tables 510 and 854
[b] 1971 SAUS table 845

How accurate were your projections? Did you take into consideration the effect of possible large price increases? Did you consider the effect of severe competition between consuming nations for a limited resource? Did you consider speculation in natural resources by underdeveloped nations with big stores of natural resources?

GAME 3

Do you think that the rate of crude oil importation will have a significant impact of food export, and hence on family food prices?

Year	Crude petroleum imported as a percentage of U.S. crude petroleum use[a]		Cost of food as a percentage of median family income[b]	
	Your projection	Data as available	Your projection	Data as available
1960	20	20	30	30
1965	23	23	25	25
1968	29	29	22	22
1970	34	34	20	20
1973				
1974				
1975				
1976				
1977				
1978				
1979				
1980				

[a] 1971 SAUS table 782
[b] 1971 SAUS tables 500, 534, 535, 537

Before long, this tradeoff will be well understood by most U.S. families. For which commodity are they most anxious to keep prices down, gasoline or food?

GAME 4

Project the effect that rising food prices will have on the value of farm-land relative to the value of other land.

Year	Food cost as a percentage of median family income		Value of all U.S. private farmland as a percentage of the value of all U.S. private non-farmland[a]	
	Your projection	Data as available	Your projection	Data as available
1952	27	27	69	69
1960	21	21	39	39
1965	18	18	37	37
1968	15	15	36	36
1972				
1973				
1974				
1975				
1976				
1977				
1978				
1979				

[a]1971 SAUS table 524

Notice that in the past, the value of agricultural land relative to non-agricultural land has dropped as food has taken a smaller bite out of family incomes. Do you think that will continue in the future? If food prices rise, on the other hand, what will that do to the value of farmland relative to urban land? What are the implications for the rate at which farmland is converted to city use at the urban perimeter? Will U.S. cities stop growing, and become high-density places, as in Europe and Asia?

GAME 5

Do you think there will be any relation between food prices (one index of inflation) and birth rates in the future?

Year	Food cost as a percentage of median family income		Number of births per 1000 women 20–24 years of age	
	Your projection	Data as available	Your projection	Data as available
1960	21	21	258	258
1965	18	18	197	197
1968	15	15	167	167
1972				
1973				
1974				
1975				
1976				
1977				
1978				
1979				
1980				

[a] 1971 SAUS table 60

In the past, food has never been expensive enough to have any effect on birth rates: other factors have been overriding. However, if inflation in general, and food costs in particular continue to soar, this may have a depressing effect on the birth intentions of young couples. Thus, birth rates might continue to drop. What will that do to you, if you teach school or university, sell toys, phonograph records, athletic equipment, skis, motorcycles, bicycles, miniskirts, or electric guitars?

By now you have the idea. It is fun to make guesses about the future. If you are a decision maker, the accuracy of such guessing can have a large effect on your personal success. To guess accurately, you have to cultivate a very comprehensive, "systems" oriented view of the future. Now try to invent more such games for yourself, which demand thought about the dynamics of the system in sectors of the economy, or society, of particular interest to you.

APPENDICES

APPENDIX I. SOCIAL REGULATION BY NEGATIVE FEEDBACK CONTROL

This appendix describes how a social problem can be countered by replacing the positive mechanisms causing or inducing the problem with negative mechanisms that can reduce, alleviate, or even solve the problem. The problem discussed is traffic congestion. The theory developed here for its solution, however, may be applied in other social and environmental problem areas as well.

First discussed is how the positive feedback system is currently operating in typical city-freeway interactions. In other words, why freeways themselves generate a greater need for freeways. Also discussed is the city-freeway-suburb interaction. While a city has an effect on the character of the transportation system, the transportation system has an effect on both the city and its suburbs. Secondly, we will describe the self-regulating mechanisms inherent in negative feedback systems, and further, how these systems might be institutionalized, replacing the positive systems.

Suppose that traffic congestion in an intraurban freeway system becomes so severe that on weekdays between 7:00 AM and 9:00 AM and again between 4:30 PM and 6:30 PM the average speed of traffic is less than 10 miles per hour. Three alternate solutions to this mess are available: (1) build more freeways, (2) encourage car pools and bus riding, and (3) build a mass transit system.

Solution 1 operates as a positive feedback loop. A city that solves a freeway congestion problem by adding to the freeway network positively reinforces freeway development for years into the future. As a consequence, the city expands, the volume of traffic increases, congestion is worse than ever, and still more freeways are required. The end point of this system occurs when a high proportion of space is covered by freeways, and rush hour congestion is so great that average traffic speeds are under 5 miles per hour.

Why does this happen? How does this affect the character of a city and its urban fringe? There are two essential components to explain why freeway traffic congestion exacerbates itself.

First, by building more freeways it is possible for people to "flee to the suburbs" at night after work. Initially it is possible to commute a fairly great distance because of the freeway speeds that are possible on new freeways before they become saturated. A given number of urban residents are now able to take up a much greater land area than they would in a compact city. Thus, freeways encourage urban sprawl. Since the distance between points (for example, home and office) becomes greater in areas of urban sprawl, more square feet of freeway or roadway are required per commuter. Freeways are self-defeating: the more you have, the more you need.

The second component of this explanation is that each car in a population of commuting cars takes up a lot more space per occupant than a bus or train per occupant. Common sense and simple observation make this point, as illustrated in Table 1.

This illustration is even more dramatic if we compare these right-of-way space requirements after correcting for the higher average speed of trains. Since trains have velocities about five times greater than cars or buses, we will use .11 rather than .55 for our space requirement per person for trains.

$$\text{buses to cars} = \frac{.10}{2.3} = .04; \qquad \text{trains to cars} = \frac{.11}{2.3} = .05$$

In this illustration cars take up twenty-five times as much space

TABLE 1

(The car-lengths of space include half the space in front of and behind
the vehicle in rush hour traffic.)

Vehicle	Car-lengths of space required on right-of-way vehicle in rush hour traffic	People carried per vehicle in rush hour	Car-lengths of space required on right-of-way per person
car	3	1.3	2.3
bus	4	40.0	.10
train (eight coaches)	220	400.0	.55

per person as buses, and twenty times as much space as trains.

Solution 1, freeway additions, creates positive feedback because each new freeway increases the square footage of right-of-way required per person in the population. This is further intensified by the fact that right-of-way requirements are higher for cars than any other form of transportation. On the other hand, solutions 2 and 3, buses (and carpools) and train systems, create negative feedback. Because they require much smaller amounts of right-of-way space per person, these methods do not create the need to build still more transit facilities to relieve still more congestion. Aesthetically, mass transit facilities are more pleasing because they are quieter, pollute less, use fewer resources, and interfere less with living. Transit systems are safer and they usually cost less per capita over the long run.

How can society replace the positive feedback system which stimulates freeway construction with negative feedback systems designed to reduce space requirements and resource demand?

Legislation should be promoted which would divert state gas tax revenues to mass transit construction according to some formula based on need. Average rush hour freeway traffic speeds should be calculated in each urban area in order to determine which proportion of gas tax revenues are to be used for mass transit. I suggest that revenue diversion occur in accordance with the formula in Table 2.

TABLE 2

Mean city traffic speed in freeways at rush hour (in miles per hour)	Proportion of state gas tax revenues to be diverted to mass transit
50	none
40	.07
30	.19
20	.35
15	.47
10	.63
8	.72
4	1.00 (all diverted)
0	1.00 (all diverted)

If any legislature wishes to be more precise, the formula used in computing the above table is: Percentage diverted = 155 − 40 × the natural logarithm of speed.

Various methods of promoting car pools and bus riding are possible and should be initiated (or improved). This is an especially good intermediary action for congested areas because existing resources are used more efficiently. In urban areas with smog problems there is an even greater incentive to take meaningful actions. I strongly support the inverse toll system. Tolls should be collected at practical locations where freeways enter urban centers. Many places already have toll collection systems which can be utilized in this plan. See Table 3 as an example.

TABLE 3

Occupancy	Toll charge per vehicle
Car with driver only	$1.25
Car with driver and one passenger	$1.00
Car with driver and two passengers	$. 75
Car with driver and three passengers	$ .50
Car with driver and four passengers	$. 25
Buses, or mass transit trains	no charge

APPENDIX II. A METHOD FOR MAKING MORE EFFICIENT USE OF ENERGY IN TRANSPORTATION SYSTEMS

It will be clear to most people that when freeway traffic surpasses some threshold, people could be better moved by buses than cars. Also, at some still higher traffic volume, buses should be replaced by mass transit trains. It would be useful to municipal, regional and state planners to have some objective means of deciding the particular traffic volumes at which these replacements should occur.

Suppose one assumes that an overriding factor in the near future will be the energy cost per passenger mile of transportation. Then for any given traffic volume, the best mode of transportation would be that giving the lowest energy cost per passenger mile.

I have not found enough information in any one place to construct a system for making these decisions. However, I have been able to gather enough data from a number of different sources to indicate a method, and give a preliminary, approximate notion of the traffic volumes and energy consumption data in the accompanying graph. In this chart energy consumption is expressed in terms of gallons of gasoline consumed per passenger mile; traffic flow is expressed in persons travelling per foot width of right of way per hour. The figures for cars and buses assume freeway widths (roadways are 60 or more feet wide).

Careful research should be initiated by the federal government to produce a graph of this type based on adequate measurements. However, such research would probably reveal a similar pattern to that in this chart.

Particularly given the prospect of rapidly rising fuel prices from now on, it would appear reasonable to convert from cars to buses, and when traffic volumes suffice, buses to trains, as indicated by such graphs.

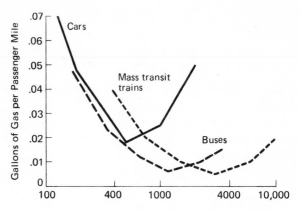

A method for deciding the most efficient mode of passenger transit for given traffic volumes.

APPENDIX III. CITIZENS' ACTION
ORGANIZATIONS

Below is a selected list of nationally active groups. The most complete list is periodically compiled by the National Wildlife Federation and sold for $2 under the title *Conservation Directory.* Upwards of 1000 American and Canadian listings are found here for government and private groups organized at international, national, and local levels whose stated objectives are related to conservation of natural resources. Very recently (in 1973), the Sierra Club Special Publications office released the first *World Directory of Environmental Organizations.* Describing a total of 1,644 organizations in nearly 200 countries, it is useful for both groups and individuals as the environmental movement becomes a world-wide force.

CALIFORNIA TOMORROW, Monadnock Building, 681 Market Street, San Francisco, Calif. 94105. Actually more of a local group, but of national interest due to its innovative planning literature and conferences. A non-profit educational organization fostering awareness of California conservation and planning problems. Publishes *Cry California* quarterly.

COMMITTEE AGAINST GOVERNMENT SECRECY, P. O. Box 4995, Washington, D.C. 20008. Formed in 1973, this is one of the newest organizations which has stated objectives to reform government. As has been illustrated, secrecy and lack of disclosure in government functions counteracts reform efforts in all areas of social change, including environmental ones.

COMMITTEE FOR ENVIRONMENTAL INFORMATION, 438 N. Skinker Blvd., St. Louis, Mo. 63130. Membership largely from the scientific community; tax deductible; grant and gift supported. Publishes information about the effects of technology on the environment. Published *Environment,* a monthly magazine.

Since the formation of CEI in 1958, similar groups have been organized in other cities with the aim of providing unbiased scientific information relevant to a variety of public issues. The Scientists' Institute for Public Information, 30 E. 68th Street, New York, N.Y. 10021, provides leadership for these groups and has adopted *Environment* as an official publication.

COMMON CAUSE, 2030 M Street NW, Washington, D.C. 20036. Volunteers and staff lobby for national reforms. Effective, aggressive, and membership directed. Has regional offices. The 1973 legislative priorities included: campaign finance disclosure, conflict of interest, open meetings, lobbying regulation, and tax credit for campaign contributions. Members receive the monthly newsletter *Common Cause Report from Washington*.

CONSERVATION FOUNDATION, 1717 Massachusetts Avenue NW, Washington, D.C. 20036. Privately supported organization for research, information, education. Publishes the *CF Letter* monthly.

DEFENDERS OF WILDLIFE, 1346 Connecticut Avenue NW, Washington, D.C. 20036. A non-profit educational organization, dedicated to preservation of all forms of wildlife.

ENVIRONMENTAL ACTION, INC., Room 731, 1346 Connecticut Avenue NW, Washington, D.C. 20036. Legal, political and social change in a broad range of environmental issues. Grew out of the national organization efforts for Earth Day 1970. Publishes *Environmental Action,* a monthly news and fact letter.

ENVIRONMENTAL DEFENSE FUND, 162 Old Town Road, East Setauket, N.Y. 11733. A non-profit organization of lawyers and scientists taking legal action in many areas, such as power and energy, land use, water resources, pesticides, highways, wildlife, and health. With 43,000 supporting members, and 700 members in the Scientists Advisory Committee. Branch offices in Washington, D.C. and Berkeley. Publishes the *EDF Letter*.

ENVIRONMENTAL LAW INSTITUTE, Suite 614, 1346 Connecticut Avenue NW, Washington, D.C. 20036. Conducts analytic and investigative research, publishes materials which further develop environmental law. Publishes *The Environmental Law Reporter*.

FRIENDS OF THE EARTH, 529 Commercial Street, San Francisco, Calif. 94111. Aggressive conservation organization founded by David Brower, membership supported. Regional and international representatives (from affiliated organizations). Especially interested in energy issues. Publications include a variety of book collections (action and natural in orientation), and the monthly magazine and newsletter *Not Man Apart*.

GARDEN CLUBS OF AMERICA, 598 Madison Avenue, New York, N.Y. 10022. The national organization of local garden clubs. Promotes horticulture and landscaping, and natural resource conservation. Offers scholarships, educational material, and information on relevant legislation.

IZAAK WALTON LEAGUE OF AMERICA, 1800 North Kent Street, Suite 86, Arlington, Va. 22209. With 56,000 members organized in many chapter or state divisions, the League promotes means and opportunity to educate the public to conserve, maintain, protect, and restore the soil, forest, water, and other natural resources of the United States.

LEAGUE OF CONSERVATION VOTERS, 1730 M Street NW, Washington, D.C. 20036. A non-partisan political organization attempting to affect government by promoting the election of public officials who will work for a healthy environment. Evaluates Congressmen's voting records and publishes *How Your Congressman Voted on Critical Environmental Issues.*

LEAGUE OF WOMEN VOTERS, 1730 M Street NW, Washington, D.C. 20036 Another non-partisan organization with 160,000 members working to promote political responsibility on a variety of fronts, including air and water conservation. Active participation stemming from Chapter and State organizations affects government at all levels. Many books and reports are available; publishes *The National Voter* monthly.

THE NATIONAL AUDUBON SOCIETY, 950 Third Avenue, New York, N.Y. 10022. Founded in 1905, this is among the oldest and largest of conservation organizations nationally. Local chapter, state divisions, issue and other committees, and staff are organized to lobby, educate, take legal actions, plan outings, gather statistics (i.e. the annual Christmas bird count), and to publish literature, including a beautifully illustrated magazine titled *Audubon.*

NATIONAL WILDLIFE FEDERATION, 1412 16th Street NW, Washington, D.C. 20036. Organized in 1936 to create and encourage an awareness for the need of wise use and proper management of the nation's natural resources. Offers study grants, information, books and gifts. Publications include the *National Wildlife Magazine* (bimonthly), *Conservation News* (semimonthly), and the *Conservation Directory* (annual).

NATURAL RESOURCES DEFENSE COUNCIL, 1710 N Street NW, Washington, D.C. 20036. Sponsors or co-sponsors litigation on behalf of the natural environment. Also have a New York Office.

THE NATURE CONSERVANCY, Suite 800, 1800 N. Kent Street, Arlington, Va. 22209. Dedicated to the preservation of natural areas. Purchases or otherwise acquires lands for science and educational purposes. Since 1958, and with the help of 26,000 members, hundreds of parcels have thus been purchased. Publishes *Nature Conservancy News.*

PLANNED PARENTHOOD—WORLD POPULATION, 810 Seventh Avenue, New York, N.Y. 10019. A federated non-profit health agency joining nearly 200 affiliates which maintain medically supervised clinics providing family planning services and information. Volunteers used in clinics for staff, organizing, and educational services.

PUBLIC INTEREST LAW GROUP, 2000 P Street NW, Room 511, Washington, D.C. 20036. A non-profit public interest law firm and headquarters for similar student funded groups organized in several states, all of which are engaged in research, investigation, and litigation of a variety of issues, including environmental ones. Directed by Ralph Nader.

RESOURCES FOR THE FUTURE, 1755 Massachusetts Avenue NW, Washington, D.C. 20036. This group undertakes research and educational programs and has been funded since 1952 by outside non-profit institutions. Books, reports, and abundant statistical information are available on American natural resources.

SIERRA CLUB, 1050 Mills Tower, San Francisco, Calif. 94104. Founded in 1892 and currently around 140,000 members strong, the Sierra Club works in the United States and other countries to restore the quality of the natural environment and to maintain the integrity of ecosystems. With nine regional offices, hundreds of local groups or chapters, and dozens of special program or issue committees, a diversified program is carried out in public education, environmental and political action, and legal actions. These are complemented by programs to "study, explore, and enjoy wildlands." Information fact sheets, several series of book publications, film lending services, and library services can be utilized by members and the public. Monthly publication of the *Sierra Club Bulletin* and chapter newsletters.

ZERO POPULATION GROWTH, INC., 4080 Fabian Way, Palo Alto, Calif. 94303. Since 1968 over 25,000 people have joined one of 250 ZPG chapters organized in order to achieve the goal of a stabilized population by 1990 by voluntary means. Educational, legal, and legislative advocacy are among the activities. Paul Ehrlich is the Honorary President. Publishes issue-oriented publication entitled *Equilibrium* quarterly, and the *National Reporter* monthly.

REFERENCES AND NOTES

Frequent reference is made throughout the book to tables from the *Statistical Abstract of the United States,* published each year by the Bureau of the Census, Social and Economic Statistics Administration, United States Department of Commerce. In reference to such tables, the abbreviation SAUS will be used.

CHAPTER 1. THE *TITANIC* EFFECT

1. Associated Press dispatch on report from the U.S. Labor Department's Bureau of Labor Statistics.

2. The reasoning behind the computation is as follows:
   What is the difference in the time required to deplete a resource depending on whether use continues at the present rate, or continues rising exponentially? In the first case, if the current use is X per year, and the time required to deplete under this rate would be Y years, then the present total amount of the resource remaining is YX.

   In the second case, use each year grows in accord with the following time series, where X is the use next year, and r is the rate by which that use increases each year, and e is the base of natural logarithms:
   $X, Xe^r, Xe^{2r}, Xe^{3r}, Xe^{4r}$, etc.
   The sum of this series is given by

$$\int_{t=0}^{t=n} Xe^{rt}dt = X\frac{(e^{rn} - e^{r \cdot 0})}{r} = \frac{X}{r}(e^{rn} - 1) \tag{1}$$

   in which t stands for the number of the year, and n is the length of time in years for which the series lasts. We wish to solve for n, the number of years a resource would last, as a function of Y, the time it would last under a con-

253

stant use schedule, and r, the rate at which the resource use would increase under exponentially increasing use. Equating (1) and YX, we have

$$\frac{X}{r}(e^{rn} - 1) = YX,$$

or

$$e^{rn} = rY + 1$$

or

$$n = \frac{\log_e(rY + 1)}{r}$$

CHAPTER 2. THE ENERGY CRISIS

1. Collated from various sources, including *Historical Statistics of the United States*, U.S. Department of Commerce, and the SAUS. M. King Hubbert, in "The Energy Resources of the Earth" (*Scientific American*, September, 1971), noted that cumulative world production of crude oil by the end of 1969 was 227 billion barrels. Since 17 billion barrels were used in 1970, cumulative use at the end of 1970 would be 244 billion barrels.

2. In *Resources and Man*, San Francisco: W. H. Freeman and Company, 1969.

3. Here the exponential growth model has been applied to a graph by Preston Cloud in William W. Murdoch (ed.), *Environment*, Stamford, Conn.: Sinauer Associates, 1971 (p. 73).

4. From SAUS for 1969, Table 1259; SAUS for 1972, Table 1333.

5. K. E. F. Watt and Peter J. Hunter, "Projecting Global Energy Demand," *Petroleum Engineer*, February, 1972 (pp. EM6–EM7).

6. SAUS for 1972, Table 1095.

7. SAUS for 1972, Table 1077.

8. *Resources and Man*, Figure 6.5 (p. 127).

9. SAUS for 1972, Tables 820 and 821.

10. SAUS for 1972, Table 822.

11. Exponential assumption applied to data in SAUS for 1972, Table 820.

12. (150,000 trillion B.T.U./3224 trillion B.T.U.) × 73 thousand megawatts = 3396 thousand megawatts (73 thousand megawatts calculated from 387 × 18.8 per cent).

13. *Current Status and Future Technical and Economic Potential of Light Water Reactors, March, 1968:* a document prepared for the U.S. Atomic Energy Commission, Division of Reactor Development and Technology.

14. SAUS for 1972, Table 825.

15. *Wall Street Journal,* November 14, 1972.

16. Associated Press dispatch, March 1, 1972, quoting Wilber H. Mack, chairman of a committee study group representing six petroleum and pipeline companies.

17. SAUS for 1971, Table 832.

18. Kenneth E. F. Watt, *Principles of Environmental Science,* New York: McGraw-Hill Book Company, 1973, Table 6–2 (p. 155).

19. R. J. Smeed, *The Traffic Problem in Towns,* Manchester Statistical Society, 1961 (pp. 1–59).

CHAPTER 3. THE RISING PRICE OF FOOD

1. Data from SAUS for 1972, Table 1316.

2. Data from SAUS for 1972, Table 1010.

3. United Press dispatch, December 2, 1972.

4. Statement by Carroll Brunthaver, Assistant Secretary of Agriculture, December 27, 1972. Quoted in *U. S. News and World Report,* January 8, 1973.

5. C. T. deWit, "Photosynthesis: Its Relationship to Overpopulation," in A. San Pietro, F. Grew and T. J. Army (eds.), *Harvesting The Sun,* New York: Academic Press, 1967 (pp. 315–20).

6. Council on Environmental Quality, *First Annual Report to the President,* 1970.

7. deWit, "Photosynthesis" (see note 5).

8. Production Yearbooks, Food and Agriculture Organization of the United Nations.

9. *The World Food Problem; A Report of the President's Science Advisory Committee,* Volume 3, Washington: The White House, 1967.

10. The number in the first row is from Colin Clark, *Population Growth and Land Use,* London: Macmillan, 1967 (p. 153). Those in the first column are from *The World Food Problem* (p. 138). The conversion of 3,600 calories per kilogram of cereal food comes from tables of nutritional data.

11. Data from SAUS for 1972, Tables 1288 and 1292.

12. Data from SAUS for 1972, Table 1066. This is the 1971 yield per acre of wheat, from SAUS for 1972, Table 1010.

13. Reid A. Bryson, "Climatic modification by air pollution, in Nicholas Polunin (ed.), *The Environmental Future,* New York: Macmillan, 1972 (pp. 134–77).

14. Gordon Manley, "Temperature trends in England, 1698–1957," *Archiv. f. Met. Geophys. und Biokl.,* B., 9:413–33. Vienna, 1958.

15. Compiled from Evans and Ruffy's *Farmer's Journal,* 1814–18.

16. This projection can be arrived at in several ways, either by using computer simulation models based on state-of-the-art climatological theory (unpublished M. S. thesis by Al Epes, Department of Mechanical Engineering,

University of California, Davis), or by extrapolating statistical time series (See Bryson reference, note 13).

17. K. E. F. Watt, *Principles of Environmental Science,* New York: McGraw-Hill, 1973 (p. 280).

18. W. E. Ricker, "Food from the Sea," in *Resources and Man,* San Francisco: W. H. Freeman, 1969 (pp. 87–108).

CHAPTER 4. ENVIRONMENTAL POLLUTION

1. See, for example, L. B. Lave and E. P. Seskin, "Air Pollution and Human Health," *Science,* 169:723–33 (1970); R. J. Hickey, "Air Pollution," in W. W. Murdoch (ed.) *Environment,* Stamford, Conn.: Sinauer Associates, 1971 (pp. 189–212); W. Winkelstein, S. Kantor, E. W. Davis, C. S. Maneri and W. E. Mosher, "The Relationship of Air Pollution and Economic Status to Total Mortality and Selected Respiratory System Mortality in Men," *Archives of Environmental Health,* 14:162–71 (1967).

2. *Proceedings of the Sixth Berkeley Symposium on Mathematical Statistics and Probability: Effects of Pollution on Health,* ed. L. M. Lecam, J. Neyman and E. L. Scott, Berkeley and Los Angeles: University of California Press, 1972. See in particular the papers by J. R. Goldsmith and J. Neyman.

3. Data for these tables are taken from *Vital Statistics of California 1965–1966–1967* and *Vital Statistics of California 1968,* published by the California Department of Public Health. In all cases the most recent figures available were used.

4. Hickey, "Air Pollution" (See note 1).

5. Based on population statistics from the 1970 census, and from the California State Department of Public Health, Bureau of Health Intelligence.

6. W. Winkelstein *et al.,* "Air Pollution and Economic Status" (see note 1).

7. SAUS for 1972, Table 79.

8. Lave and Seskin; also Hickey (see note 1).

9. Reid A. Bryson, "Climatic Modification by Air Pollution" in Nicholas Polunin (ed.), *The Environmental Future,* New York: Macmillan, 1972 (pp. 133–77).

10. See, for example, San Francisco *Sunday Examiner and Chronicle,* November 28, 1971, section A, p. 26.

CHAPTER 5. MARKET SATURATION AND SLOWDOWN

1. Data from SAUS for 1972, Table 1227.

2. From SAUS for 1971, Table 327.

3. Data from SAUS for 1971, Table 345.

4. Calculated from data in SAUS for 1971, Table 1240.

5. W. Stewart Pinkerton, Jr., in the *Wall Street Journal*, March 6, 1970.

6. SAUS for 1971, Table 889.

7. From statements by Stuart G. Tipton, president of the Air Transport Association, as reported by the Associated Press on June 15, 1970, and by United Press International on February 3, 1971.

8. This is an estimate often made in interviews by airline and aerospace executives. See, for example, the *Wall Street Journal*, June 6, 1972, September 8, 1972, and April 6, 1971; also the Los Angeles *Times*, April 25, 1971.

9. See, for example, an Associated Press dispatch quoting Haughton in the Sacramento *Bee*, Dec. 15, 1971; also an article by Senator Henry Jackson in the San Francisco *Sunday Examiner and Chronicle*, February 1, 1970.

10. Quoted in the Sacramento *Bee*, May 14, 1971.

11. One example is an article by Joe Alex Morris, Jr., in the Los Angeles *Times*, December 19, 1971.

12. *Wall Street Journal*, July 31, 1972.

13. *Wall Street Journal*, October 6, 1972.

14. Data collated from the Los Angeles *Times*, December 19, 1971, and *Flight International*, 16 November 1972.

15. Calculated from SAUS for 1972, Tables 932 and 1340; cross checked against data in *Interavia* (an international aviation magazine), and *Flight International* (both for 1972).

16. From SAUS for 1972, Table 936.

17. Data from *Flight International*, November 16, 1972, cross checked against data from *Interavia* and newspaper accounts of sales.

18. SAUS for 1971, Tables 895, 896.

19. Associated Press dispatch, San Francisco *Examiner*, February 6, 1972.

20. SAUS for 1972, Table 932.

21. SAUS for 1971, Table 872 (railroads); Table 890 (airlines).

22. Calculated from SAUS for 1972, Tables 2, 932.

23. *Business Conditions Digest*, U.S. Department of Commerce.

24. From an article by Paul E. Steiger, Los Angeles *Times*, May 2, 1971.

25. See note 8, above.

26. Report by the Bank of Hawaii, *Sunday Star-Bulletin and Advertiser*, Honolulu, November 21, 1971.

27. Jack Miller, San Francisco *Examiner–Chronicle*, March 7, 1971.

CHAPTER 6. INFLATION

1. News item, *Wall Street Journal*, April 23, 1973; article by Walter W. Heller, *ibid.*

2. Associated Press dispatch, December 15, 1972; *U. S. News and World Report,* January 8, 1973.

3. SAUS for 1971, Table 990.

4. Commodity commentary issued by Dupont Glore Forgan. Also daily newspaper price quotations.

5. SAUS for 1971, Table 1024.

6. Preston Cloud, "Mineral Resources in Fact and Fancy," in William W. Murdoch (ed.), *Environment: Resources, Pollution and Society,* Stamford, Conn.: Sinauer Associates, 1971 (pp. 71–88).

7. Daily quotations in newspaper financial sections.

8. United Press International dispatch, in *Rocky Mountain News,* Denver, June 15, 1972.

9. SAUS for 1972, Table 632.

10. SAUS for 1971, Tables 632 and 634, with data from Table 2.

11. John Kenneth Galbraith, *The New Industrial State,* Boston: Houghton Mifflin, 1967 (Chapters 16 and 24).

12. *Wall Street Journal.*

13. SAUS for 1970, Table 325.

14. SAUS for 1970, Table 325.

15. Honolulu *Advertiser,* July 8, 1971. Both reports originated with United Press International.

16. Boyce Rensberger, *The New York Times;* reprinted in the Sacramento *Bee,* September 20, 1972.

CHAPTER 7. A COMING GLUT OF MANPOWER

1. Data from SAUS for 1972, Table 8.

2. Data from SAUS for 1972, Table 8.

3. Data from SAUS for 1972, Table 340.

4. Simon S. Kuznets, *Economic Change: Selected Essays In Business Cycles, National Income, and Economic Growth,* New York: Norton, 1953.

5. Richard A. Easterlin, *Population, Labor Force, and Long Swings In Economic Growth: The American Experience,* New York: Columbia University Press, 1968.

6. Los Angeles *Times,* February 21, 1971.

7. San Francisco *Sunday Examiner and Chronicle,* November 28, 1971.

8. SAUS for 1972, Table 367.

9. SAUS for 1972, Table 65.

10. Frederick C. Klein and Richard D. James, *Wall Street Journal,* March 8, 1971.

11. Data from SAUS for 1971, Table 890; also SAUS for 1972, Tables 932 and 933.

12. SAUS for 1969, Table 317.

13. SAUS for 1972, Table 820.

14. SAUS for 1970, Table 351.

15. SAUS for 1970, Table 941.

16. SAUS for 1972, Table 369.

17. SAUS for 1972, Table 340.

18. SAUS for 1972, Table 508.

19. SAUS for 1970, Table 338.

20. SAUS for 1972, Table 341.

21. Calculated from SAUS for 1972, Table 7.

22. Calculated from SAUS for 1972, Table 340, as follows:
    For 1971, unemployed = 4.993 million
    For 1970, unemployed = 4.088 million
    $$\overline{}$$
    905 thousand

23. Report of a press conference by John B. Connally in Los Angeles *Times,* January 30, 1972.

24. Los Angeles *Times,* January 30, 1972.

25. Data from Bureau of Labor Statistics.

26. SAUS for 1970, Table 817.

27. SAUS for 1971, Table 810.

CHAPTER 8. INTERNATIONAL MONETARY TURBULENCE

1. Data from SAUS for 1972, Tables 1288 and 1292.

2. Data from SAUS for 1972, Tables 1288 and 1292.

3. *Wall Street Journal,* September 21, 1972.

4. *Wall Street Journal,* May 11, 1972.

5. *International Herald Tribune,* Paris, June 28, 1972.

CHAPTER 9. WRONG PRIORITIES, MISALLOCATED RESOURCES

1. *U. S. News and World Report,* September 11, 1972, p. 46. The BART system will cost about $1.4 billion for 75 miles. The total cost of urban and suburban mass transit systems would be 80 million people × 50 miles of track per million people × $20 million per mile of track = $80 billion.

2. *U. S. News and World Report,* February 21, 1972, p. 37.

3. Jack Miller, San Francisco *Examiner–Chronicle,* March 7, 1971.

4. *Sunday Star-Bulletin and Advertiser,* Honolulu, November 21, 1971, p. F–12.

5. See note 3.

6. *Sunday Star–Bulletin and Advertiser,* Honolulu, July 11, 1971, p. A–21.

7. SAUS for 1972, Table 932.

8. *Wall Street Journal,* November 14, 1972.

9. John Kenneth Galbraith, *The Affluent Society,* Boston: Houghton Mifflin, 1958.

10. Frederick C. Klein and Richard D. James, *Wall Street Journal,* March 8, 1971.

11. Rexford G. Tugwell, *A Model Constitution for a United Republics of America.* James E. Freel and Associates, 577 College Avenue, Palo Alto: 1970.

CHAPTER 10. COMMITMENT TO ECONOMIC GROWTH

1. SAUS for 1972, Table 889.

2. SAUS for 1972, Table 1230.

CHAPTER 11. OBSESSION WITH PRODUCTION INSTEAD OF SERVICES

1. The figures in this table were selected or calculated from SAUS for 1972, Tables 1333, 1322, 1318, 1321.

2. Figures from *Historical Statistics of the United States from Colonial Times to 1957,* Bureau of the Census, U.S. Department of Commerce, 1960; and SAUS for 1972, Table 820.

3. Herman E. Daly, ed., *Towards a Steady-State Economy,* San Francisco: W. H. Freeman and Co., 1973 (pp. 160–68).

CHAPTER 12. OVER-POPULATION

1. SAUS for 1971, Tables 2 and 614.

2. Norman R. Glass, K. E. F. Watt and Theodore C. Foin, "Human Ecology and Educational Crisis: One Aspect of the Social Cost of an Expanding Population," in S. F. Singer (ed.), *Is There An Optimum Level of Population?,* New York: McGraw-Hill 1971 (pp. 205–18).

3. Data from various Statistical Abstracts (see, e.g., SAUS for 1969, Table 7, and SAUS for 1972, Table 7).

4. Gross national product growth rates computed from SAUS for 1972, Table 506.

5. Simon S. Kuznets, 1953, *Economic Change: Selected Essays in Business Cycles, National Income, and Economic Growth,* New York: Norton, 1953; also Richard A. Easterlin, *Population, Labor Force, and Long Swings in Economic Growth; The American Experience,* New York: Columbia University Press, 1968.

6. Calculated from SAUS for 1971, Tables 632, 634, and 20.

7. SAUS for 1972, Table 22.

CHAPTER 13. THE CONVENTIONAL WISDOM

1. SAUS for 1972, Tables 1235 and 557.

2. *Wall Street Journal,* September 26, 1973; also September 27, 1973 (chart of Dow Jones averages)

CHAPTER 14. SCENARIOS OF THE FUTURE

1. *Historical Statistics of the United States from Colonial Times to 1957,* Washington: Bureau of the Census, U.S. Department of Commerce, 1960, Table M71–87.

2. This is the high estimate given by M. K. Hubbert in *Environment,* ed. William W. Murdoch, Stamford, Conn.: Sinauer Associates, 1971 (p. 104).

3. *Ibid.,* pp. 104–105.

INDEX